Rough and Tumble

# Rough and Tumble

Aggression, Hunting,
and Human Evolution

## Travis Rayne Pickering

**UNIVERSITY OF CALIFORNIA PRESS**
Berkeley · Los Angeles · London

University of California Press, one of the most
distinguished university presses in the United States,
enriches lives around the world by advancing
scholarship in the humanities, social sciences, and
natural sciences. Its activities are supported by the UC
Press Foundation and by philanthropic contributions
from individuals and institutions. For more information,
visit www.ucpress.edu.

University of California Press
Berkeley and Los Angeles, California

University of California Press, Ltd.
London, England

Library of Congress Cataloging-in-Publication Data

Pickering, Travis Rayne.
    Rough and tumble : aggression, hunting, and human
evolution / Travis Rayne Pickering.
        p. cm.
    Includes bibliographical references and index.
        ISBN 978-0-520-27400-6 (cloth : alk. paper)
        1. Hunting, Prehistoric. 2. Hunting and gathering
societies. 3. Fossil hominids. 4. Human evolution.
5. Social evolution. I. Title.
    GN799.H84P53 2013
    599.9--dc23
2012027494

Manufactured in the United States of America

22  21  20  19  18  17  16  15  14  13

10  9  8  7  6  5  4  3  2  1

In keeping with a commitment to support environmen-
tally responsible and sustainable printing practices, UC
Press has printed this book on Rolland Enviro100, a
100% post-consumer fiber paper that is FSC certified,
deinked, processed chlorine-free, and manufactured with
renewable biogas energy. It is acid-free and EcoLogo
certified.

*For whom I love the most*
*Anneliese and Grace—the reason*
*Evelyn, Bob, and Lance—the source*

# Contents

# Illustrations

# Acknowledgments

I procrastinated and fretted over this brief testament of appreciation. Having to arrange my gratitude always induces in me a certain overwhelming queasiness. And how many different ways can I say "Thank you"? I'm not sure, but I do know that all the simple variations of that sentiment that I employ here cannot express sufficiently my appreciation to those to whom I am so mightily indebted.

With that in mind, my first words of thanks are to Craig Stanford, who gave me the initial advice on how to get this project published. Without that direction, this book would still be languishing in digital limbo on an obsolete PC. Once taken, the path to publication was circuitous, and I thank Russ Tuttle for kind words and early advice that improved the first incarnation of the book, as well as Michael Fisher, editor at Harvard University Press, for giving the manuscript a chance and then recommending that I get in touch with Deirdre Mullane. Deirdre has not only proven a wonderful literary agent, but also a minor miracle worker, coaxing a (short!) full-length book out of this writer, who I am sure is her most reticent client. Blake Edgar, senior sponsoring editor at University of California Press, and editorial coordinator, Kate Marshall, have been great in shepherding me through the publication process; many thanks to them both, as well as to an anonymous reviewer, the UC Press editorial committee, and the anonymous presenter of the committee's decision on my book. Since he waived anonymity, I also thank Trent Holliday by name for his helpful and complimentary review of the manuscript. Very special thanks to Pam Suwinsky, my excellent copy editor. Finally, I'll be honest. It is not without pressing down the envy of a frustrated artist that I am able to thank and congratulate Dan Bunn for executing the painting that graces this book's dust jacket.

I wrote this book while holding down a day job. Thanks to all my coworkers at the University of Wisconsin-Madison, especially Henry Bunn, Jill Capps, Mary Griese, John Hawks, Clara Pfefferkorn, Karen Steudel, and Karen Strier, for their vocational support and camaraderie. In South Africa, it's the Swartkrans Paleoanthropological Research Project team members, Bob Brain, Ron Clarke, Jason Heaton, Kathy Kuman, Andrew Phaswana, and Morris Sutton, whom I thank, as well Francis Thackeray and Evlyn Ho (at the Institute for Human Evolution, University of the Witwatersrand), Lazarus Kgasi, Tersia Perregil, and Stephany Potze (at the Ditsong National Museum of Natural History), Bernhard Zipfel (at the Bernard Price Institute for Palaeontological Research, University of the Witwatersrand), and Andrea Leenan (at the Palaeontological Scientific Trust). Thanks also to everyone who is part of the Olduvai Paleoanthropology and Paleoecology Project (Tanzania), helmed by Manuel Domínguez-Rodrigo, Enrique Baquedano, Henry Bunn, and Audax Mabulla.

Several colleagues, including some whom I don't know very well and others who are old friends, were much more than generous with their time, encouragement, and advice. I am doubly grateful to some of them for their permissions to use the images included in this book. Alphabetically, these colleagues are Rebeca Barba, Lee Berger, Bob Brain, Henry Bunn, Steve Churchill, Ron Clarke, Nick Conard, Paul Constantino, Christopher Dean, Manuel Domínguez-Rodrigo, Korrina Duffy, Brian Hare, Jason Heaton, Job Kibii, Kathy Kuman, Bruce Latimer, Isaac Makhele, Colin Menter, John Mitani, Jacopo Moggi-Cecchi, Nkwane Molefe, Abel Molepolle, Jim Moore, Stephen Motsumi, Jill Pruetz, Mohamed Sahnouni, Kathy Schick, Kara Schroepfer, Sileshi Semaw, Jordi Serangeli, Pat Shipman, Neil Shubin, Jingzhi Tan, Nick Toth, Erik Trinkaus, Alan Walker, David Watts, Tim White, Bernard Wood, José Yravedra, and Zhang Yue. Without the long-term, unconditional support of five of these people—Bob, Henry, Manuel, Ron, and Kathy (Kuman)—this book could not have been conceived or written. At the risk of effusiveness, I think that each of you knows of my fondness for you and what you mean to me as a mentor and confidant. More important, I hope that I demonstrate that fondness to you adequately and always; please know how very grateful I am to you. The same goes to three other close friends, Jason Heaton, Charles Egeland, and Morris Sutton. And, while I'm being sentimental, I also extend my very deepest thanks to the whole Brain/Newman/Watson and Clarke/Kuman clans, and to Shuna Huffman, for all being like family to me and for making South Africa my home away from home.

As for my biological family, this book's dedication is my life distilled to its essence: wife, daughter, mom, dad, and brother. No one and nothing could be more important to me than are all of you. I love each and every one of you more than I can put into words. Thanks, in turn, for loving and supporting me. (Biscuit and Stella, you're not so bad either; I'm much obliged for your unrelenting insistence to take a break already and go for a walk. And, you Linda Harrison?—well, you're so outstanding I hardly consider you a mother-in-law at all.)

Last, this book was completed in spite of Flannery O'Connor, whose ghost every day drawls to me in her voice sweet as honey, "Everywhere I go I'm asked if I think the universities stifle writers. My opinion is that they don't stifle enough of them."

# Introduction

*Homo homini lupus est.*
(Man is a wolf to man.)
—Plautus, *Asinaria*

Human nature is perhaps best defined by its malleability, and we can be sure that aggression is a component of our versatile character. If today's behavioral scientist refers to "innate drives," he or she is considered anachronistic. Nonetheless, we still recognize a myriad of internal, biological bases (both structural and biochemical) for aggressive expression in humans and in other animals—even when that aggression is prompted by external stimuli. The capacity for human aggression, including lethal aggression, must have been shaped by evolutionary forces. In forthcoming pages, I discuss some well-known hypotheses of the evolutionary basis of human aggressive potential. By way of introduction, I mention here the view of political anthropologist Christopher Boehm, who makes a cogent point regarding aggression and its relationship to human nature. Boehm argues that

> when Darwinian competition becomes direct and face to face, as it tends to be in highly social animals such as humans, chimpanzees, wild dogs, or scores of other species, it is precisely dispositions to dominance (producing threat and attack—[that is, aggression]) and submission (producing appeasement or flight) that are retained by natural selection as useful behavioral strategies. Like most any other behavior trait, these reproductively selfish political dispositions are maintained through individual maximization of inclusive fitness [that is, the evolutionary success of an individual's genes as measured by the number of those genes he *and his relatives* pass on to survive in subsequent generations].

Recruiting an abundance of ethnographic data, Boehm goes on to demonstrate the evolutionary and political usefulness of aggressive capableness for maintaining egalitarianism in simple human societies, in which the threat of physically antagonistic (including mortal) retaliation against a tyrant by the collective submissive is an ever-real potentiality. Thus, in this view, the "fixed" inherent aggressive *potential* of humans is a very useful trait, indeed. Recognizing the circumstantial utility of aggression is not the same as arguing that humans are fundamentally aggressive. Instead, it is the simple acknowledgment that aggression is a behavioral tactic—one among many along a broad spectrum of human behavioral potential—that can be recruited to satisfy (or, at least, attempt to satisfy) the innate human dispositions to dominate and to resist being dominated.

Boehm's outlook on the utility of aggressive potential in the sociopolitical setting is perfectly reasonable, but here I argue that the aggressive potential of humans is a double-edged sword. I posit that aggressive potential was also a capacity often to be constrained by our ancestors. My argument about restricting aggression is quite apart from any moral judgment about the application of interpersonal violence. That is an issue for ethicists and, indeed, in real time, for the individual moral human actor. And, although a relevant evolutionary aspect of tamed aggression is that it can yield fitness recompense by avoiding physically damaging conflict with competitors, my argument does not concentrate on this point either. Instead, the idea I forward, while acknowledging that aggressive attack is a perfectly efficient way for our chimpanzee cousins to kill prey (nourishing and energizing them in order that they can then pursue mating opportunities, which lead, in turn, to increased Darwinian fitness), also posits that the same kind of hunting would have been a hopeless tactic for early human hunters, who, in comparison to their large, potentially dangerous prey, were small, weak, and slow footed.

The old anthropological tradition of coupling hostile propensities and hunting prowess needs to be abandoned, and human aggression and human hunting delinked. That delinking will be accomplished only by our deeper appreciation of how primitive weapons might have mediated hunting by our earliest ancestors. Decoupling ideas of aggression and our understandings of human predation and human evolution in no way nullifies the importance of our coming to grips with violence and venality. In many ways these are facets of the human condition that might stand beyond the purview of science. They are forces that stay bound, unbreakably, to the human species, infrangible across the 6 million-plus years of our evolution and in the face of all of our civilizing efforts over the millennia. Joseph Conrad expressed it plainly when he commented that man is always, at all times—before, today, and after—"quite capable of every wickedness." But, foisting this awareness onto models of our emergent past does nothing to illuminate the elusive transition from ape to human. Like others before me, I argue that hunting was a primary factor in our becoming fully human—a factor underpinning the completely unique ways in which we organize ourselves and interact with others of our own kind. This means, in turn, that we need to characterize human predation as accurately as possible in order to build the fullest and most realistic understanding of what it is to be human.

# 1

# A Man among Apes

Make it thy business to know thyself, which is the most
difficult lesson in the world.

—Miguel de Cervantes, *based on the ancient Greek aphorism,*
  *"Know thyself"*

I have a remarkable friend named Bob Brain. In his early eighties, Bob
now finds himself among the last of the true natural historians. Not
much worth knowing about the complex ecological interplay of organism and environment escapes his deliberation. Lately, Bob's mind is on
the origins of animal life. The particulars of his work on that topic are
beyond the bounds of this book, but an overview is relevant. Although
Bob's research is rooted in theory, it's driven by the recovery of fossils;
in any historical science, data generated in the course of well-conceived
fieldwork are the definitive sources of testable hypotheses. And, Bob's
observations are defying conventional wisdom about the first appearance of the animals. He and his colleagues are documenting spongelike
organisms fossilized in Namibian limestones. These fossils are in excess
of 750 million years old, significantly more ancient than customary estimates of when the first animals emerged from their simpler eukaryotic
ancestors. These results do not, however, inspire in Bob the same kind
of rapture that has motivated other heretic intellectuals in their own
battles with establishment "big boys." Instead, Bob's probing of the
beginnings of multicellularity only confirm his long-held convictions
that humankind in this new century stands at the brink of its own
destruction, and that we tilt toward that abyss as the ultimate
consequence of more than half a billion years of heterotrophy. Prior
to the animals, all organisms were autotrophic, using inorganic carbon dioxide to fulfill their energetic requirements. The evolution of

heterotrophy changed all that, and those organisms that entered this intensely competitive system began to ingest other life forms in order to satisfy their carbon needs. The food web was spun, and consumerism, in the basest sense of that term, was born. One evening, overlooking a hard southern African sunset, Bob contemplated this natural state of affairs and deadpanned to me that "once these bastards evolved there was no turning back. It's quite disgusting, really."

It is from this basic biological perspective, and with an appreciation of the great depth of the Earth's geological history, that an uncharitable reader could dismiss the contents of this book. Is predation not now, more than 700 million years after its emergence, rendered just as mundane as it is profound? Animals kill, animals eat. Humans are animals, so they kill and eat. So what? But, consider just how unique an activity is human hunting. It's true that some nonhuman animals use tools to hunt, others target prey larger than an individual hunter, and a few even share food—but none, other than the human animal, possesses these traits as a *behavioral complex* used in combination to both satisfy their caloric and nutritional requirements *and* to build social cohesion. Many researchers argue that this type of multifaceted, socially complex hunting—still employed today by the last of the traditional human foragers—is the very socioecological basis of our humanness. If so, then human predation is undoubtedly a topic worthy of serious scholarship. Indeed, it has already generated an enormous body of research during the 150-plus years since the study of human evolution was first codified as paleoanthropology, a true, empirically based scientific endeavor. This book draws on that research, situating notions of human predation within a generalized paleoanthropological framework.

In addition to well-reasoned hypotheses of human evolution, part of that framework is the sociohistorical milieu in which those hypotheses were produced. Paleoanthropology never operated in a vacuum. Global events, concerns of science at large, and the proclivities of its practitioners conspired time and again to connect ideas about human predation to studies of aggression. And, invariably, new data from the human fossil record or from research on great apes, our closest living relatives, eventually called for a reassessment of that purported linkage between hunting and hostility.

Currently, we are firmly in the latter part of this repeating cycle. In the past fifteen or so years, it is primatologists who have generated the most provocative ideas about the presumed connections between human aggression and hunting and about how our earliest ancestors interacted with their biological symbionts and competitors. In particular, the keen observations of extant primates by Craig Stanford, Richard Wrangham, and Donna Hart and Robert Sussman have led to prominent but divergent conclusions about human origins. Drawing on his fieldwork in the forests of Tanzania, Stanford argued that primordial men were, like chimpanzees, eager and efficient killers, flesh-hungry *Hunting Apes*. Viewing the world through a darker lens, Wrangham upped the ante. His broad survey of ape behavior suggested to him that our forebears were not just hunters but that our agnatic ancestors were *Demonic Males,* hyperantagonistic louts, fully capable of dragging women around by their hair, and worse. Just as the collective work of Stanford and Wrangham brought the blood-letting to a fever pitch by the early 2000s, Hart and Sussman entered the scene to offset any such notions of Stone Age machismo and brutishness. They reminded us that most wild primates live a precarious existence, surrounded by and subject to the whims of hungry, prowling predators. Arguing by analogy, they continued that humans of the past must surely have faced the same looming conditions as do the primates of today. This opposing construct gave us *Man the Hunted,* the idea that our ancestors lived under continual menace from sabertooth cats, giant hyenas, and even large birds—eking out a spare existence in the shadows of a predator's world. So it is that we are left with the latest incarnations of the two great, contrasting narratives of human evolution: early man as misanthrope and mighty hunter versus early man as milksop.

In my opinion, there is merit in both of these views of early human life. However, I also argue that in perhaps underestimating—and surely underreporting—the importance of a rich archaeological record, produced by our Stone Age ancestors and stretching back at least two and a half million years into the past, the otherwise excellent accounts of my monkey- and ape-studying colleagues lack *the* essential component of the story. Simply put, the archaeological record stands as testament to the *actual* behavior of our prehistoric forerunners. Admittedly, this stone-and-bone witness of past action is accessible to us only in glimpses—the record is woefully incomplete, subject to the vagaries of ancient preservation and modern discovery. But those glimpses are still imbued with a genuineness of testimony that data on modern primate behavior can only approximate in their stead. Paleoanthropologist Tim White has opined in a similar context, "The rich detail of the modern world compared to the paucity of the prehistoric world can serve to obscure the recognition and analysis of evolutionary novelty. The present illuminates the past in myriad ways. However, the unwary paleobiologist can easily misinterpret past organisms by using inappropriate interpretive constructs based solely on modern form and function."

My already mild critique of the primate-centric approach to reconstructing our evolutionary history is softened even more by the undeniable fact that the work of Stanford, Wrangham, Hart, and Sussman (as well as many other primate specialists) is as fundamental as it is exceptional. In fact, without it science lacks the proper comparative framework to discern those precious few aspects of human uniqueness that might also be the basis of *humanness*. Those two things—human uniqueness and humanness—are, of course, different. Chimpanzees don't build skyscrapers, and baboons can't conduct orchestral symphonies (although it would be fun to watch them give it a go). More than that, though: neither a chimpanzee nor a baboon could even create and manage a simple campfire without a human's prompting and training. But these uniquely human capabilities—grand and humble—are, all the same, just overlays on a now very, very deeply contained human essence, which, if to be revealed, will require the efforts of not just geneticists, psychologists, and primate behaviorists but also those of archaeologists and paleoanthropologists. This little book concerns itself primarily with presenting current data from these latter two disciplines in supplement to the fine primatological work already extant. Combining these datasets is intended to uncover the basis of humanness from natural history records that are always tantalizing but that are also usually imperfect and, oftentimes, frustratingly resistant. Reduced, the story is a simple one. Human hunting underlies humanness. Successful human hunting is necessarily decoupled from human aggression. Tools, in enhancing the distance between a human hunter and its nonhuman prey, facilitated this decoupling of action and emotion.

The first proposition isn't original. Many paleoanthropologists have, over a long time, argued that hunting underpins humanness. Language is the only other proposed prime mover of "becoming human" that equals the influence of the "hunting hypothesis." Determining the prehistoric emergence of either—human-styled hunting or language— remains elusive, but that does not discourage continuing efforts. Most of those intellectual labors pivot around *Homo erectus,* a well-known species of extinct human ancestor that existed between 1.8 million and 500,000 years ago.

## DOES THE BOY MAKE THE (HU)MAN?

The discovery, relegation, and eventual scientific acceptance of *Homo erectus*—and the psychological impact of its transition through those stages on its discoverer—is often told. In the end, Eugène Dubois emerges as a flawed hero, one whose emotional oscillation between impracticable tenacity and crushing resignation is finally vindicated posthumously, and only in historical retrospection. The story began valiantly enough in 1887 when Dubois bucked the shared academic wisdom that humans first evolved in Europe. Undeterred by archaic-looking (and thus possibly quite ancient) Neandertal fossils already known from various locales in Europe, Dubois acted instead on Darwin's prescient notion about the biogeography of human origins. Like Lord Monboddo (James Burnett), the famous Scottish jurist of nearly one hundred years before him, Darwin recognized that because our closest living relatives, the apes, are all tropical species, then the tropics must be where our most recent common ancestor with the apes resided, as well as where the earliest members of each descending lineage must have evolved. From that elegantly reasoned starting point, a three-year stretch of Dubois's search for man's earliest ancestor was, in contrast, epitomized by a paleontological naïveté surpassed only by its human cost. Supporting himself as a medical officer in the Royal Dutch East Indies Army, Dubois's fossil prospecting on Sumatra was an abject failure: no truly ancient fossils were recovered, one of his two engineers died, and several members of his conscripted prisoner work crew deserted. Others of those who toughed it out were wretchedly ill for much of the expedition. It was only in 1890 that Dubois finally moved to more fertile ground in Java, where a human skull discovered a few years earlier piqued his interest. By November 1890, Dubois's crew had unearthed a gnarly piece of lower jawbone that was undoubtedly that of a hominin (that is, a member of the zoological group that includes modern humans and all extinct species that are more closely related to us than they are to chimpanzees, with whom the hominins shared a most recent common ancestor about 6 million years ago). Motivated, Dubois continued his searches in Java, where, in October 1891, he found a hominin skullcap on the banks of the Solo River, and a hominin thigh bone ten months later. This was a truly impressive haul, especially considering that Dubois pressed on in spite of the orthodoxy of the time that largely consigned human origins research in the tropics as a wild goose chase.

Unfortunately, even when presented with Dubois's impressive proof to the contrary, that prevailing attitude did not desist. The best reception Dubois found for his fossils was lukewarm. Based on its morphology, competent anatomists could not deny that the thigh bone would have supported a two-legged, upright-walking (bipedal) hominin. But, most also refused to accept an association between the thigh bone and the beetle-browed skullcap, which they conjectured belonged to a giant, gibbonlike ape rather than to a transitional human species, as Dubois claimed. Initially, Dubois rallied against this poor treatment and defended the hominin status of *Pithecanthropus erectus* ("upright ape-man," the original scientific moniker of *Homo erectus*) on the European lecture circuit, but around the turn of the twentieth century he abruptly silenced himself on the subject and hid the fossils.

Dubois's interpretation of *Homo erectus* as a genuine human ancestor was ultimately vindicated by subsequent discovery of the species's remains at other sites throughout Asia, Europe, and Africa. Study of those finds gradually built up a compelling picture of an animal that was quite distinct from *Australopithecus,* the genus of true ape-man species from the Pliocene and lower Pleistocene (geological epochs that spanned, collectively, 5.3 million–780,000 years ago), and from the very earliest putative hominins like *Ardipithecus,* which first appeared about 7 million years ago, late in the Miocene Epoch. Conspicuously, *Homo erectus* skulls—with brainpan volumes ranging from 600 to 1067 cubic centimeters, and with an average somewhere around 880 cubic centimeters—are much larger than those of modern apes, ancient ape-men, and putative Miocene root hominins. (Compare the relatively impressive cranial capacity of *Homo erectus* to an apish 350- to 600–cubic centimeter range for *Australopithecus,* 300–350 for *Ardipithecus,* and a modern human average of about 1400 cubic centimeters.) Corresponding to the expanded braincase of *Homo erectus,* is its reduced, somewhat less projecting face as compared to apes and earlier occurring hominins. Inferred from isolated scraps of the skeleton, sketchy estimates put *Homo erectus* adults at around five to five and a half feet in height and about 120 pounds. Each of these findings was an exciting incremental advance beyond Dubois's rudimentary understanding of *Homo erectus.* But, it was a discovery made nearly thirty years ago in the fossil-rich badlands of northern Kenya that truly pulled back the veil to reveal *Homo erectus* in its emergent humanness. The first miniscule fragments of what would eventually be pieced together into a nearly complete skeleton of *Homo erectus* (figure 1) were found

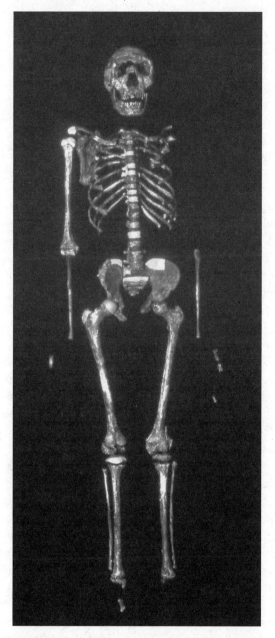

FIGURE 1. The 1.5 million-year-old skeleton of Nariokotome Boy, from northern Kenya. The child, when he died, was well on his way to adulthood; likewise, his skeleton represents an early hominin species, *Homo erectus*, that was well on its way to becoming human. (Photograph courtesy of Alan Walker)

by legendary fossil hunter Kamoya Kimeu at an unassuming site called Nariokotome.

Analysis of the Nariokotome *Homo erectus* skeleton disclosed a seemingly young boy with a body of nearly adult stature—a body that was long and linear, the kind of physique that is best adapted to the tropics, able to maximize heat dissipation across its relatively expansive surface area compared to its relatively small volume. Moreover, Nariokotome Boy's arms and legs are also proportioned in the same way as are those of modern people. This implies to many anatomists that he possessed a more efficient, modern humanlike upright gait than did the putative root hominins and ape-men, some of which had extremely wide hips, relatively short legs, and even opposable big toes—morphology that would have, in comparison to *Homo erectus,* cost these types of hominins some efficiency in two-legged, bipedal striding. The narrow but deep, barrel-shaped ribcage of Nariokotome Boy is also like ours and contrasts with the inverted funnel-shaped ribcages of apes (more on this seemingly innocuous difference later in this chapter).

Initially, it seemed that the boy had died well before he was done growing. In order to estimate the boy's age at death, researchers used their knowledge of the ages at which different parts of modern human skeletons stop growing. First, all of Nariokotome Boy's permanent teeth, except his upper canines and upper and lower wisdom teeth, are erupted: in terms of the pace at which the adult teeth of modern humans develop and emerge into the mouth, this condition places the boy between ten and ten and a half years old when he died. In contrast, the elbow end of the boy's upper arm bone, the humerus, had just begun to fuse (it is not until a long bone is done growing that the joint surfaces at both ends fuse permanently to the shaft in the middle) when he died, giving an estimated age of around thirteen years old at death. Probably based on his large body, most researchers favored the older estimate, derived from the data on his long bone fusion, as the most likely age at which the boy died. And this was where the situation stalled for several years.

But more recently, Christopher Dean and Holly Smith, specialists on early hominin teeth, surprised many paleoanthropologists by concluding that Nariokotome Boy may have instead been as young as eight years old when he died. In order to ascertain the time elapsed in tooth development at his death, Dean and Smith studied the boy's perikymata, growth bands on tooth enamel that form incrementally at particular rates. Complicating Dean and Smith's task was the fact that the Nariokotome

Boy died when he was an older juvenile, meaning that most of his baby teeth were already shed and several of his erupted permanent teeth had already ceased developing. Thus, no single tooth preserved in his jaws records time for the whole of his short life (as can be the case for another mammal that died as a younger juvenile). To overcome this difficulty, Dean and Smith started counting perikymata in the earliest formed tooth that is available in Nariokotome Boy's jaws. They then continued by picking up the uninterrupted count in another tooth that overlapped with the first tooth but that had also continued growing beyond the completed formation of that first tooth. In addition, because a tooth's roots continue growing for some time after its crown is completely formed, Dean and Smith needed to add estimated elapsed time of subsequent root growth to the Nariokotome Boy's teeth with completed crowns.

Another hitch in most perikymata studies is that the number of days between the formation of any tooth's adjacent perikymata, or periodicity, varies among individuals, although for modern people average periodicity is eight days. Nariokotome Boy, however, has only a few perikymata on his front teeth; research has shown that teeth with few, widely spaced perikymata usually have higher than usual periodicities (in other words, greater than eight-day periodicities). In addition, the state of Nariokotome Boy's tooth root formation argues against a typical eight-day periodicity for him. For these reasons, Dean and Smith suggest that a ten-day periodicity is more likely to characterize Nariokotome Boy, which, combined with the estimates of tooth root formation, places him between almost eight and a half and nearly nine years old when he died. (Using instead the typical modern human eight-day periodicity makes Nariokotome Boy even younger at death, between seven and a half and eight years old.)

Momentum in this mini-renaissance of our understanding of *Homo erectus* biology accelerated in 2008, when added to Dean and Smith's startling findings was the description of a newly discovered female *Homo erectus* pelvis from a fossil site in Ethiopia called Busidima. Analysis of the Busidima *Homo erectus* pelvis shed further light on just what kind of animals were Nariokotome Boy and the rest of his species. The bony birth canal of the Busidima pelvis would have accommodated delivery of a baby with a brain volume of around 315 cubic centimeters, 30 to 50 percent of the adult brain size reconstructed for *Homo erectus*. In this respect, *Homo erectus* was like a modern person, with impressive prenatal brain development. But, because the adult *Homo erectus* brain reached (at best) only two-thirds of the mass of a modern human's, it means that its remaining postnatal brain growth was completed relatively more quickly after birth than it is for humans. The upshot is that *Homo erectus* children—even those as young as the Nariokotome Boy—were probably highly independent and physically capable much earlier in life than are modern human kids. Even allowing for his long, lean build at the time of his death, the initial extrapolations of a six-foot-one-inch to six-foot-five-inch adult into whom a supposed thirteen-year-old Narioktome Boy would have grown was always perplexing. Dean and Smith's case for an eight-year-old Nariokotome Boy, *already nearing his adult height and weight when he died,* is less mystifying but holds just as profound—if different—implications.

Taken together, the new growth data for Nariokotome Boy and the Busidima *Homo erectus* pelvis paint a fascinating picture of a truly transitional animal—one that was like modern apes (and extinct root hominins and ape-men) in having a brief childhood, but also like modern humans in having significant brain growth *in utero*. In other words, Nariokotome Boy (and presumably all *Homo erectus* individuals, if he is indeed representative of the species) at eight years old was as physically mature as an ape of the same age but far exceeded the corporeal development of a modern eight-year-old human child, instead approaching more closely that of a fifteen-year-old boy. So, what does this mean for *Homo erectus* if we pose the crass (and probably unanswerable) question: Did the species, for all its admitted impressiveness, manage to traverse that indistinct threshold across which an ancestral hominin became a human?

In superlative irony, it is the Nariokotome Boy's principal analyst who has been most ardent in answering no to that question. It was Alan Walker, an eclectic intellectual with a background in geology and primate anatomy, who assembled and led the team of scientists that first studied Nariokotome Boy's skeletal biology. Walker's colleague, Ann MacLarnon, analyzed the skeleton's thoracic vertebrae and concluded, because of the constricted central canals of those vertebrae, that the boy possessed a spinal cord that would have been smaller in his thoracic region than that of a modern person. Specifically, based on comparisons with the differential distribution of spinal cord tissues in modern primates, Nariokotome Boy would have had a smaller amount of spinal cord gray matter in his thoracic spinal column than do modern people. This, in turn, indicated to MacLarnon that people living today have greater innervation of their thoraxes than did *Homo erectus*. By extension, modern people also have greater control over their breathing than did *Homo erectus*. Because precise control of breathing is a requisite for human speech, Walker and MacLarnon contend further that *Homo erectus* was unable to talk. And, for Walker, "At some deep level, being fully human is predicated upon being linguate. That meant that the boy my colleagues and I spent so many years discovering and analyzing was profoundly *in*-human."

Thus for those who equate language and humanness, the case on the humanness of *Homo erectus* was assumed closed: Nariokotome Boy lacked language, therefore he was not human. More recently, however, paleoanthropologists Bruce Latimer and Jim Ohman argued that the narrow thoracic vertebral canal of Nariokotome Boy is aberrant, its constriction not from normal development of the boy but instead a manifestation of pathological axial dysplasia. For Latimer and Ohman this means that we should not assume that the structure of Nariokotome Boy's vertebral canal is typical of *Homo erectus* as a species—and, as follows, conclusions about language deficiencies in the species based on the boy's skeletal morphology are, at best, premature. MacLarnon and her collaborator Gwen Hewitt countered that a pathologically narrowed vertebral canal in Nariokotome Boy—by up to 40 percent assuming that other, "normal" *Homo erectus* individuals had canals of the same sizes as do modern humans—would have disrupted neurological communication between the child's brain and his legs. They go on to note that Nariokotome Boy's leg bones are thick and robust—evidence of their normal, habitual use, which implies further that they were, necessarily, also normally innervated.

So, if the jury is still out on the "language-makes-human hypothesis" as it articulates with what we know about *Homo erectus,* then what about "man the hunter"—or "woman the gatherer," or "the home base/male-female food-sharing model," or any of those other once much vaunted hypotheses of humanness that I discuss in the forthcoming pages? For now, it is sufficient to note that each was proved ultimately a caricature of human cultural evolution, but also that within the core of that scrap heap is a nugget worthy of excavation. The starting premise of each of those discarded antiquations—that social organization is a practical marker of humanness—remains a pragmatic one. That is because the social organization of our ancient ancestors is potentially *detectable* in the archaeological and fossil records.

## THE GUT OF THE MATTER

Modern hunter-gatherers are not untouched by the machinations of bordering agropastoralists and ever-encroaching industrialized societies, but they still maintain a cultural status among living people that most closely approximates the "natural" social condition of *Homo sapiens*. What are the circumstances that allowed for this fundamental state of organization to develop from an apelike existence, and when did it occur: at only 200,000 years ago, when modern humans first appeared; before that, with the emergence of *Homo erectus* at 1.8 million years ago; deeper in time, with the beginning of the genus *Homo* at roughly 2.3 million years ago; or, even before, with the australopithecines or Miocene root hominins? Time and again, that question has been answered by invoking a single "magic catalyst." And each such answer is ultimately (if not immediately) annihilated.

An increased regularity of meat eating (beyond that observed in modern apes) *does* seem to be an essential component of the basic human subsistence strategy, but it is only part of a complex feedback system. Katharine Milton, a specialist on primate diets and nutrition, elaborated on this nuanced awareness: "It is the behavioral trajectory taken by humans to secure high-quality foods—rather than simply the foods themselves—that has made humans human. . . . Evolving humans appear to have relied increasingly on brain power as the key element in their dietary strategy, using technological and social innovations to secure and process foods before ingestion." Human foraging societies developed out of this evolutionary strategy. And so, unsurprisingly, humanization was a process, rather than, as "magic catalyst" thinking evokes, a singular event.

Milton further discriminates her model from a "meat-as-the-magic-catalyst hypothesis" by viewing animal protein not as an exclusive nutritional focus of ancestral hominins, but instead as a dietary facilitator that allowed those hominins to intensify their exploitation of carbohydrate-rich plant foods. According to Milton, high-energy plants, *rather than meat,* were the most important fuels supporting increasingly larger brains (and bodies) as the genus *Homo* evolved since it first appeared about 2.3 million years ago. Because meat is so *densely nutritional,* consuming it regularly can ease the high-cost search and recovery of glucose-laden plants—resources that are patchily distributed and seasonally oscillating in their availability. Even a relatively small amount of meat (or of other carcass resources, like skin, marrow, and brains) efficiently satisfies daily requirements for essential fatty acids and basal energy.

This fact—that, with meat, a little gets you a lot—eventually became the basis of an influential explanation for a shift in hominin body form, which was tracked through the course of human evolution. The human brain is a voracious consumer; at just around 2 percent of the body's total mass, it demands nearly 20 percent of its daily intake of energy and oxygen. Paleoanthropologists Leslie Aiello and Peter Wheeler referred to the brain as *the* most important expensive body tissue in their "expensive tissue hypothesis" of the evolution of the genus *Homo*. But, like the brain, the heart, kidneys, liver, and gastrointestinal tract are also very demanding metabolically. Aiello and Wheeler argued that it was impossible for all these other organs to be maintained at such high cost if the hominin brain was to enlarge over time; there is a limit on the amount of energy that an organism is capable of capturing every day. The problem is that an early hominin would not survive very long with compromised heart, kidney, or liver function. Properly functioning guts are also essential to life, but there is more biological latitude for variation in gut size and long-term survivorship. According to Aiello and Wheeler, this all means that reducing gut size was the only evolutionarily workable option for early *Homo* as it evolved larger brains but also continued to maintain critical life-supporting bodily functions.

Aiello and Wheeler also suggested that paleontological evidence confirmed their hypothesis. Typically, hominin soft tissues, like brains and guts, do not fossilize. However, ancillary bony evidence that *is* readily preserved can inform us indirectly about the respective sizes of our ancestors' brains and guts. The empty space that the brain occupied in life can be measured on fossilized skulls, providing a size estimate of that once-living, now-decayed organ. Comparisons of skulls from different hominin species show that brain volume increased between hominin species through time. Gut size is trickier, but probably not impossible, to pin down using the osseous fossil record. Aiello and Wheeler wrote, "The large gut of the living [apes] gives their bodies a somewhat pot-bellied appearance, lacking a discernable waist. This is because the rounded profile of the abdomen is continuous with that of the lower portion of the rib cage, which is shaped like an inverted funnel." Based on the fragments of ape-man ribs and vertebrae recovered up to the early 2000s, most paleoanthropologists inferred that the earliest members of the genus *Australopithecus* had thoraxes with the same apish morphology as the trunks of modern gorillas and chimpanzees. In addition, the configuration of *Australopithecus* hips, which are more widely flaring at their tops than is expected from the estimated heights of individual ape-men, was also used to reconstruct their guts as protuberant.

However, in 2005, a 3.6 million-year-old ape-man skeleton, from
the species *Australopithecus afarensis,* was discovered at the Ethiopian
site of Woranso-Mille. The skeleton preserves several partial ribs and a
partial pelvis, allowing reconstruction of the shape of its thorax. To the
surprise of many, the skeleton's reconstructed thorax does *not* have the
inverted funnel shape of an ape's, thought previously (based on more
fragmentary rib fossils) to also characterize the ape-men. Instead, the
Woranso-Mille ribcage is uniquely bell shaped, wide at the top, like
that of early *Homo* and modern humans—*and, wide at the bottom.*
(An even more recent study of *Australopithecus afarensis* thoracic vertebrae, recovered from the Ethiopian site of Hadar, agrees that ape-men
had ribcages that were in many ways more humanlike than was previously appreciated.) Implications of this more accurate understanding
of *Australopithecus* thorax shape for the "expensive tissue hypothesis"
remain unexplored, but I note that inferences of an expansive ape-men
gut are unchanged by the Woranso-Mille skeleton; it is only the shape
of the *top* of the specimen's ribcage—unrelated to gut size—that is
unexpected.

More direct assault on the "expensive tissue hypothesis" comes from
a recent study on brain size and organ mass in one hundred modern
mammal species, including twenty-three types of primates. Anthropologists Ana Navarrete, Carl van Schaik, and Karin Isler demonstrated that
when "controlling for fat-free body mass, brain size is not negatively
correlated with the mass of the digestive tract or any other expensive
tissue, thus refuting the expensive tissue hypothesis." The researchers
concluded that brain enlargement in hominins must have, therefore,
been enabled instead by a *combination* of biological phenomena, which
not only increased energy capture (as is possible with a higher-quality
diet) but that also reduced overall energy demands of the individual.
Among other adaptations, Navarrete and her colleagues point to the
increasing efficiency of two-legged bipedal walking over the long course
of human evolution (as evinced by more recent, larger-brained *Homo
erectus* having longer legs and narrower hips than did earlier, smaller-
brained *Australopithecus,* with its shorter legs and wider hips) as a
reduction in energy expenditure relative to the climbing and four-legged
quadrupedal locomotion of nonhuman apes.

Regardless, the large brain size and barrel-shaped ribcage of *Homo
erectus,* as exemplified by the Nariokotome Boy, mean that—no matter
how and why those morphologies evolved—evolution ultimately set up
a situation in which *Homo erectus* was now locked in an inescapable

feedback system, a system in which reliable access to animal product was no longer a luxury but a necessity. The small, short guts of *Homo erectus* did not provide long enough transit times to extract from bulk vegetation its complexly bound energy and nutrients that these hominins needed to feed their large, hungry brains. In addition, by almost 1.8 million years ago, there is incontrovertible evidence from the site of Dmanisi, in the (former Soviet) Republic of Georgia, of *Homo erectus* living outside of Africa, along the southern slopes of the Caucasus Mountains. Subsisting in that markedly seasonal environment, with its long winters during which plant food was unavailable, necessitated that *Homo erectus* was a proficient meat forager—like modern humans, using weapons to kill animals larger than an individual hunter, and, because of the large size of prey, probably sharing meat within a group. Indeed, the preponderance of archaeological evidence from *Homo erectus* sites worldwide (not just from temperate regions, like the Caucasus Mountains) supports the idea that the species as a whole had developed these humanlike abilities by as early as 1.8 million years ago.

But, can we push deeper in time? Were the ape-men also perhaps more men than apes in their predatory prowess? A puckish, early twentieth-century champion of *Australopithecus* certainly thought so. More than that, he thought he had proved it in the most dramatic way.

# Prehistoric Bloodsport

Kill the pig. Cut her throat. Spill her blood.

—William Golding, *Lord of the Flies*

And he found a new jawbone of an ass, and put forth his
hand, and took it, and slew a thousand men therewith.

—Judges 15:15, *The Bible (Authorized King James Version)*

Raymond Dart is not a household name, but twenty-five years after
his death, his concept of protohuman "killer apes" still lingers in our
collective conscious. Consumers of popular culture will recall the opening scenes of Stanley Kubrick's classic film *2001: A Space Odyssey*. The
vignette unfolds among a group of hirsute humanoids, whose building
aggression eventually becomes irrepressible when they discover that
discarded animal bones can be employed as lethal weapons. The
sequence culminates in a vivid scene of bone-bashing hyperactivity and
murder that launches the trajectory of human evolution toward our
ultimate achievement: extraterrestrial exploration. All of it—this dismal
view of the root of humanity and its innovations—derived ultimately
from the imagination of Raymond Dart. In constructing his model
of hunting australopithecines, the killer apes of our parentage, Dart
voiced his deep conviction that man's darkest nature was the key to our
evolutionary success and that of our prehistoric forebears. The germs of
this outlook developed early in Dart's career, at a time when he had just
accomplished the none-too-small feat of establishing our very point
of origin in the animal world. And his journey to that remarkable
discovery was as circuitous as it was fortuitous.

Dart was born in 1893, in Brisbane, Australia, studied science on a scholarship at the University of Queensland, and eventually qualified in medicine at the University of Sydney. His professional interest was in neuroanatomy, and after his service in the Australian Medical Corps during World War I he took jobs at various medical facilities in England and America for the better part of four years. Then, in 1921, he grudgingly accepted the position of Professor of Anatomy at the University of the Witwatersrand (Wits), in Johannesburg, South Africa.

Johannesburg, in the old Transvaal Province, was still a sinewy frontier town in the early twentieth century, and Dart considered his move there more as a penal sentence than an advance of his career. Like the American West of half a century earlier, the Transvaal was host to all manner of rough beast and man—not the least among them, the miners. The Whitewater Reef (Witwatersrand in Afrikaans) is an east–west trending range of low hills that runs across the Transvaal. The reef is the Earth's single richest source of gold, yielding nearly half of all that precious metal ever mined—a fact that is even more remarkable considering that gold was discovered there only in 1886 (including on a farm called Langlaagte, the embryo of modern-day Johannesburg). The Witwatersrand is also riddled with countless underground caves that are extremely rich in lime. Because lime is a crucial alkalizer in the Macarthur-Forrest cyanide process, used to extract gold from low-quality ore, South African gold and lime mining were intimately connected in the late 1800s and early 1900s. In order to recover lime from the underground caves, miners had to also remove (by hammering, digging, and dynamiting) breccias, the caves' calicified sediments, which interdigitated with the lime. In the process of doing so, the miners discovered that the breccias contained fossilized bones.

### THE DEEP WOMB OF HUMANITY

The ultimate story of how those bones became encased in Transvaal breccias began long before complex organisms like vertebrates had even evolved, some 2.7 billion years ago, when the Kaapvaal Craton stabilized and consolidated as one of the Earth's earliest continental masses. As the Kaapvaal Craton evolved, basins developed on it, the largest of which is the Witwatersrand Basin, which once held an inland sea and at times was also open to the ocean to its southeast. Sediments precipitated out of the sea across the basin, eventually hardening into impure dolomitic limestone.

This dolomite became the host rock for the fossil-bearing caves in the Witwatersrand Basin. The caves did not, however, develop in the dolomites until about 20 million years ago, early in the Miocene Epoch. Ancestral hominins had still not appeared by 20 million years ago, but the sea that once covered the Witwatersrand Basin had long since retreated. In deep levels under the water table, perhaps 150 or so feet below the ground surface, weakly acidic ground water dissolved calcium carbonate out of the dolomite at points of weakness, along horizontal bedding planes and vertical joints; this process formed hundreds or thousands of water-filled underground caverns of various sizes throughout the basin. Over time, the water table dropped in the basin and the caverns were now air-filled voids, but they were also still below and sealed off from the ground surface above. Rain and ground water continued to seep through cracks into the dolomite that hosted the underground caverns, all the while still dissolving calcium carbonate out of the rock. Eventually this process created vertical shafts of various sizes connecting the underground caverns to the ground surface. It is estimated that the opening of these connections coincided with the presence of human ancestors in South Africa, by the early Pliocene (5.3–3.6 million years ago).

Once open to the ground surface, various animals, including hyenas, large predatory cats, and porcupines, started to utilize the caverns for shelter and as dens. We know from observing their living descendants that all of these types of animals also happen to accumulate bones in caverns as residues from their feeding activities. The attraction of carnivores and large rodents to the underground caves is not (and was not) random. Today, the caves are nestled in a highveld ecosystem, comprised largely of open grasslands, and the Witwatersrand Basin of millions of years ago was much the same. Within this largely exposed environment, young trees tend to grow near cave openings because it is there that they can most easily take root, and because the rocky overhangs around cave openings protect them from occasional violent weather. Trees that today commonly grow around South African cave openings, such as white stinkwood and wild olive, also prefer soils rich in lime, which is exactly the type found in dolomitic caves. Further, lusher vegetation persists in this locally ideal growing environment because the caves are moist, supplying essential water (not as readily available in the well-drained open grasslands) that many trees require. The trees, in turn, attract animals seeking shelter, if even temporarily, from exposure and dangers on the open grassland.

Cagey predators do well to haunt such natural, prey-aggregating hotspots. The feeding debris from their successful kills can eventually become incorporated into the caves. Some carnivores eat their prey in the thickets above the caves, with bone residues subsequently washed underground by seasonal rains or simply falling into the chasms as the soil-covered edges of caves crumble away over time. In order to secure carcasses from ground-bound competitors (like hyenas), leopards, more specifically, will haul their kills into the trees that hang over cave openings. As a leopard feeds arboreally, bones can tumble down into the void below. In other cases, when a shaft is amenable to easy negotiation, a leopard or hyena might itself go underground and use a cavern as a feeding lair or den in which to rear cubs. Baboons are known to aggregate in large numbers in the upper chambers of caves, where they sleep together in relative warmth during cold weather (much of South Africa experiences a temperate climate, with freezing nights during the winter months; caves maintain a consistent temperature of about 50 degrees Fahrenheit throughout the year, regardless of external ambient temperature). A leopard lying in wait in a deeper recess within the same cave in which baboons have gathered above has a good chance for an easy meal. The unfortunate baboon victim is dragged back down into the same dank alcove from which its killer emerged, its bones poised to become fossils, even as the leopard digests its meat.

African porcupines are perhaps less dramatic, but certainly no less prolific, bone accumulators than are carnivores. Wildlife biologists disagree about the ultimate reason why porcupines collect bones and other hard objects to chew. A common anecdote in the American North Woods and Canada has porcupines chewing the handles of axes left in blocks of wood overnight. The well-accepted explanation for this behavior is that the rodents are attracted to the minerals deposited on the tools by human perspiration. Bones, likewise, contain minerals that porcupines might seek to supplement their diet. Researchers also recognize that a porcupine's incessant chewing of hard materials is essential to its survival because, like other rodents, porcupines possess ever-growing incisors. But, those commentators who then extrapolate that the need to trim their incisors is the root of porcupine bone-collecting habits defy logic in doing so. More in keeping with the way evolution works is the alternative proposal that the ever-growing incisors of rodents are an adaptation to their diet, rather than the diet being adapted to their physiology. Regardless, the osseous buildup created by porcupines in South African caves is impressive and easily diagnosed by the occurrence of their distinctive gnaw marks on bones.

Beyond mammals, birds of prey are obviously attracted to the trees around cave entrances. It is from roosts in these trees that especially owls launch and return from their attacks on various and sundry vermin. There are some parts of owls' kills, like hair and bones, that they cannot digest. At its roost, an owl will regurgitate these undigested body parts in the form of oblong masses called pellets; when the hair and other degradable matter (like the keratin that forms the sheath around a rodent's claws) decays, it is the rodent's bones that remain. Impressive buildups of pellets in various stages of disintegration are common below active roosts. Many of the South African fossil cave breccias are chock-a-block full of mice and gerbil bones, undeniable testament to the great antiquity of owl pellet-casting behavior.

Finally, accidents happen. We don't often see it, but animals can be clumsy or unlucky. Shafts connecting underground caves to the ground surface are often obscured by vegetation, forming dangerous "death traps" into which unsuspecting animals can tumble below to their deaths. And, it's not just an antelope running from a predator that is vulnerable to a death trap. Some animals actually choose to enter caves, but then, either as a result of injury, sickness or disorientation, cannot escape their confines. The climbing proclivity of animals like leopards and baboons is the very quality that can be the undoing of these athletic creatures. In fact, the taxonomic predominance of cats and primates is a striking feature of fossil assemblages from many of the Witwatersrand caves. In the absence of tooth marks on fossils of these animals—which might indicate their demise, consumption, and deposition in the cave by predators—a parsimonious way to explain their abundance is

> that [they] may have wandered into the cave[s] intentionally and were then unable to escape.... [The abundance of cats and primates]—taxa that are more agile than are [antelopes]—is suggestive that the [caves] were accessible, but not easily accessible or escapable, to motivated, proficient climbers.... Support for this idea [in one case] is that 32 percent of the total primate [sample] is composed of subadults, individuals that might have been less savvy than were adults in finding their ways back out of the cave. Finally, we note that "intentional entry" need not always mean an animal scampering down a near-vertical [shaft].... [Many caves have horizontal tunnels ending in precipitous drops, unapparent in the darkness.]

Bones derived from all these processes—processes that are observable today and are inferred to have also operated in the past—are eventually covered over with the dirt and rocks that wash and collapse regularly into the cavern, from the ground surface above. Under the right conditions, this complex mixture is then eventually sealed and hardened by water, charged with calcium carbonate, which drips onto it from stalactites attached to the cave ceiling above, ultimately forming a fossil-bearing breccia.

## THE FAMILY TREE

Blocks of such fossil-bearing breccias came into Raymond Dart's hands after he solicited his students for anatomical specimens to fill the new teaching museum he had established at Wits. Because they contained primate fossils, sediment blocks from Taung, a remote cave on the edge of the Kalahari Desert, were of particular interest to Dart. One of the Taung primate fossils, discovered in late 1924, is the skull of a young-ster, perhaps four years old when he died. After careful study, Dart determined that the specimen is no mere monkey, but instead represents an extinct species of human ancestor he called *Australopithecus afri-canus*—the southern ape of Africa. Although the Taung Child's skull is in many ways apelike, with a small braincase and large jaws, a key hominin trait is discernible in the base of its cranium. The hole through which the spinal cord travels to articulate to the brain, the foramen magnum, is situated underneath and toward the front of the baby's skull—the morphology of a two-legged, bipedally walking primate, whose head, like ours, rested on top of an upright spine.

In comparison to modern humans, *Australopithecus africanus*, which we now know dates from 3 to 2 million years ago, seems quite primi-tive. But, research since Dart's days has revealed several other hominin species that predate *Australopithecus africanus* by millions of years. Biomolecular evidence indicates that the hominins diverged from the chimpanzee lineage somewhere in Africa between 8 and 4 million years ago (probably closest to 6 million years ago). And, recent paleontologi-cal research across that continent is augmenting the fossil record of this critical time span.

Currently, there are three possible candidates for the oldest known Miocene hominin species. One of these, *Ardipithecus kadabba*, was discovered by paleoanthropologist Yohannes Haile-Selassie and his colleagues in geological deposits that occur in the Middle Awash River Valley of Ethiopia, and that are dated between 5.7 and 5.2 million years old. Ten years earlier, in 1994, a research group led by Tim White announced the first known species of *Ardipithecus, Ardipithecus ramidus,* which was found in the same region as *Ardipithecus kadabba* but in a geological context about one million years younger. White, Haile-Selassie, and their coworkers argue that *Ardipithecus kadabba* and *Ardipithecus ramidus* are time-successive chronospecies, which represent the earliest appearance of the hominin lineage, as *kadabba* evolved gradually into *ramidus* over time. *Ardipithecus ramidus* is the better known of the two species, represented at a site called Aramis by just over one hundred fossils that come from at least sixteen individuals, including the astonishing partial skeleton of an adult female nicknamed Ardi. (A few *Ardipithecus ramidus* fossils are also known from 4.5 million-year-old sediments at the nearby Ethiopian locality of Gona.) Reconstruction of Ardi's skeleton is a testament to the patience and paleontological skills of White and his collaborators. It was discovered in horrible condition, its bones so crushed and fragmented that before they could be removed from their matrix they had to first be hardened chemically. Next, the bones were extracted from the ground still surrounded by their encasing sediments. Covering the sediment-encased bones with additional protective layers of plaster allowed the fossils to be safely transported to a museum laboratory in Addis Ababa, where only then were they removed from the dirt under hypercontrolled conditions. Last came the protracted matter of physical and virtual (using technologies like microcomputed tomography scanning) restoration.

Like the Taung Child and modern humans, Ardi's cranium shows a foramen magnum that is positioned underneath and toward its front. Ardi was obviously a biped when she was on the ground, a fact confirmed by analysis of her pelvis. Her ilium—that upper, bladelike part of the pelvis beginning at the waistline—is short and squat, and flares sideward, like a modern human's and unlike that of a quadrupedal monkey or ape. This pelvic morphology indicates that Ardi's butt muscles were situated much like those of a person and allowed her to maintain her center of mass over her body's midline, so that her upright stride was stable and did not shift from side to side as she walked forward. In contrast, the lower part of Ardi's pelvis is like an ape's, with the bony architecture to support large leg muscles that would have been useful for climbing in trees. When in the trees, Ardi did not, however, move like apes, who brachiate, swinging underneath horizontally projecting branches. Ardi instead has hands with flexible wrist, palm, and finger joints. She also has feet with divergent and opposable big toes and stable, rigid soles (Ardi's upright walking would, thus, have been accomplished without the aid of the big toe in propulsion—meaning the other toes would have been quite powerful in pushing the body forward). This combination of hand and feet features is very different than the body plan of African apes, who have long, curved fingers and stiff hand joints to support their massive bodies not only when they climb, hang, and brachiate in trees, but also when they walk on their knuckles, ambling across the ground. African apes are also unlike Ardi in having impressively flexible feet that can be bent inward toward each other at incredible angles, so that they can maintain tight contact with the vertically rising trunks of trees as they climb up into forest canopies. In trees, Ardi would have moved more like most living monkeys, clambering along the tops of branches on the palms of her hands and soles of her feet. Ardi's thumbs are slightly shorter relative to her other fingers than are those of other hominins, but the index through pinky fingers of Ardi are not exaggeratedly elongated as in brachiating apes. More generally, Ardi is small bodied—estimated to have been about four feet tall and weighing around 110 pounds when alive—and small brained, with an estimated brain volume on par with modern chimpanzees, between 300 and 350 cubic centimeters.

*Ardipithecus* teeth are an interesting departure from the typical human condition of relatively large, thickly enameled premolars and molars. Although the enamel on *Ardipithecus* teeth is thin, it is not as thin as that of chimpanzees, who specialize in eating ripe fruits; *Ardipithecus* must have been more omnivorous than are chimpanzees, perhaps feeding on vegetation both in the trees and on the ground. But, the most significant way in which the dentition of *Ardipithecus* differs from that of an ape is in the front of the mouth, where the upper and lower canines come into contact. The humanlike form of the canines is the only feature beyond upright walking that links *Ardipithecus* more closely to other hominins than to other types of apes. Although *Ardipithecus* canine teeth are relatively large, they are also blunt and less projecting and dagger-like than are the canines of apes. In addition, apes have lower third premolars—those teeth directly behind the lower canines—that slope backward, creating gaps between them and the lower canines, so that when the jaws are closed, the upper canines lock between the lower third premolars and the lower canines. This arrangement means that, as an ape opens and shuts its mouth, the back edges of its upper canines glide across the backward sloping shoulders of its lower premolars—and, in doing so, hone the back edges of the upper canines. This natural fang-sharpening system, indicated by large, interlocking upper and lower canines, is found in primate species in which males compete aggressively for opportunities to mate with females; big canines are invaluable fighting weapons, and are, in other cases, displayed in warning yawns that can circumvent physical contact between male rivals.

In contrast, this specialized upper-canine honing system is unknown in *all* hominins, living and extinct. Humans and their direct ancestors possess (or possessed) small, stubby canines that do not or did not project significantly beyond the chewing surfaces of their other teeth. By extension, the smaller canines of hominins seem to indicate a radically different social system than observed in apes, with markedly reduced levels of male–male competition, and perhaps even cooperative intragroup behavior. Although the upper canines and lower third premolars of *Ardipithecus kadabba* (that species that preceded *Ardipithecus ramidus* in the geological record) interlocked, the structure was not so developed as to sharpen the top fangs as thoroughly and consistently as are those of apes. By analogy with what we know of the relationship between canine size and primate mating systems, it therefore seems that by nearly 6 million years ago the putative hominin *Ardipithecus kadabba* might have been organized socially in a decidedly non-ape way. This finding has implications for the central thesis that is developed in the following chapters of this book.

Before that, though, it is important to note that among critiques from other detractors, a challenge to the hominin status of *Ardipithecus* is forwarded by a Kenyan and French team of paleoanthropologists working in 6 million-year-old deposits in the Tugen Hills of Kenya. Team leaders Brigitte Senut and Martin Pickford stress the apelike features of *Ardipithecus* and relegate it to a position as an ape ancestor, while offering their own discovery, *Orrorin tugenensis*, as the ancestor of all species leading to modern humans. However, *Orrorin*, represented by various teeth, lower jawbone fragments, and pieces of arm and leg bones, is very much like *Ardipithecus* in that both animals retained many primitive features in their anatomies. The same can be said for the remarkable 7 million-year-old cranium from the Djurab Desert of northern Chad, which was recovered in 2001 and 2002 by a team led by paleontologist Michel Brunet. Some critics claim that this cranium, assigned to the novel taxon *Sahelanthropus tchadensis*, is just an ape, without definitive indications that it was an obligate biped and disqualifying it from the hominin lineage. Other researchers, such as White and Haile-Selassie, see enough anatomical continuity between *Sahelanthropus, Orrorin* and *Ardipthecus* to suggest that all three might be included in the same genus—which, if correct, would push the earliest known appearance of *Ardipithecus* one million years deeper into prehistory, based on

*Sahelanthropus,* at 7 million years old. This debate will, no doubt, continue to unfold in the always contentious and always entertaining annals of the study of the very earliest human ancestors.

Regardless, it was not until very long after the extinction of *Ardipithecus, Orrorin,* and *Sahelanthropus* that the first species of our own genus appeared, as indicated by a 2.3 million-year-old upper jawbone (with characteristic *Homo* morphology, such as a broad palate, a muzzle that projects only moderately, narrow first molars, and rhomboidal-shaped second molars) from the fossil locality of Hadar, in northeastern Ethiopia. In the 2 million-plus years between the demise of *Ardipithecus, Orrorin* and *Sahelanthropus,* sometime around 4.4 million years ago, and the rise of early *Homo,* about 2.3 million years ago, hominins were represented on Earth by the so-called ape-men, members of the genus *Australopithecus* (and some species of *Australopithecus* survived until more recently, about one million years ago, during which span they were contemporaries of early *Homo*). Dart's Taung Child, *Australopithecus africanus,* was only one of several species included in this widely distributed and long-lived group of hominins. For instance, the most famous australopithecine skeleton, Lucy, belonged to *Australopithecus afarensis,* a geologically older species known from 3.7–3.0 million-year-old sites in Ethiopia, Kenya, and Tanzania. And *Australopithecus afarensis* was preceded in East Africa by another distinct species called *Australopithecus anamensis* (4.2–3.9 million years old).

Australopithecine skeletons show a mixture of apelike and humanlike adaptations. Ape-men had small brains, ranging from a chimpish low end of around 350 cubic centimeters to an underwhelming high end around 600 cubic centimeters. The braincase of *Australopithecus* is hafted onto a large, projecting facial skeleton, reminiscent of that of a gorilla or chimpanzee. The ape-men were also small, between three and a half and five feet tall and weighing only 65 to 125 pounds when alive. For most species that are represented by male and female fossils of the same bones, we are able to infer that males were bigger than were females. But, the estimated size disparities between male and female australopithecines also vary between species. For example, the intersexual difference in *Australopithecus afarensis* body mass exceeded only slightly that known for modern humans and chimpanzees but was less than that known for living gorillas, which are characterized by enormous males who, in comparison, dwarf their females. In contrast, some analyses conclude that *Australopithecus robustus*—a South African species that was more morphologically specialized and geologically recent than was *Australopithecus afarensis*—was highly sexually dimorphic in body size, with large males whose physical development was longer and completed later in life than that of smaller females, whom they presumably dominated socially and sexually. The forearms of the ape-men are long compared to their upper arms and legs. Their hand bones indicate that they possessed powerful grips but were also capable of the fine manipulations characteristic of modern humans. Relevant cranial and skeletal features indicate that all ape-men were at least facultative bipeds; their foramen magnums are positioned under and toward the fronts of their skulls, and the morphologies of their spines, hips, thighs, and feet are adapted to varying degrees of upright walking. However, several recent finds of partial ape-men feet from across Africa demonstrate that styles of bipedalism, as well as the relative skill with which ape-men could climb trees, varied among australopithecine species. Like *Ardipithecus,* at least two species of *Australopithecus* (one represented by "Little Foot," an astonishingly complete skeleton from the South African cave site of Sterkfontein, and another from the 3.4 million-year-old Ethiopian site of Burtele) had opposable big toes—quite useful for climbing, less so for the kind of strong, confident heel strikes and big toe push-off that modern humans use to walk. In contrast, fossilized trails of ape-men footprints, preserved in solidified volcanic ash at the 3.6 million-year-old site of Laetoli, in Tanzania, indicate efficient, modern humanlike bipedalism, as does the morphology of various foot bones of *Australopithecus afarensis.*

The bipedalism of *Australopithecus* might be a bewildering mixed bag, but all species in the genus show a similar general construction of their jaws and teeth. Indeed, all ape-man species are united with one another, and are set apart from the emergent hominins of earlier times and the apes of today, by a single anatomical characteristic called *megadontia*—the enlargement of the premolars and molars, the cheek teeth that sit behind the canines. Not only did the cheek teeth of *Australopithecus* increase in absolute size through time, from species to species, but they also increased in dimension relative to body size, and they changed shape in some species, with the premolars becoming molarlike in their anatomy. In addition, the enamel of *Australopithecus* cheek teeth is much thicker than is the enamel of apes and of the earliest pre-australopithecine hominins. Collectively, these dental features functioned as part of an adaptive complex that involved increasingly more powerful chewing capabilities as the australopithecines evolved throughout the Pleistocene (figure 2).

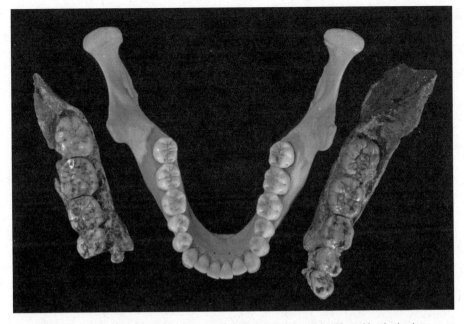

**FIGURE 2.** The complete lower jawbone of an adult modern human is bracketed by the broken halves of adult *Australopithecus* lower jawbones, illustrating the comparative enormity of ape-men teeth (the much larger size of the *Australopithecus* teeth is even more extraordinary considering that ape-men were, on average, of much smaller body size than are modern humans). The back of the jaws are at the top of the image, starting above with the third molars (or wisdom teeth) and moving down each tooth row toward the front of the mouth. (Photograph courtesy of Jason L. Heaton)

### PRECOCIOUS BEASTS WHO FLOUTED SACRED COWS

The apogee of this unique form of dentition occurs in a group of ape-men referred to informally as the "robust australopithecines." The particularities in skull anatomy that link the robust australopithecines together as a distinct subgroup of *Australopithecus* arose as developmental consequences of their extreme tooth size and a requirement for hyper-masticatory power. For example, a high ridge of bone that runs down the midline of the male robust australopithecine cranium—the sagittal crest—anchored immense chewing muscles that originated on the lower jaw. The lower jawbone itself is massively deep, and the cheek bones are placed far forward on the face of robust ape-men skulls so that their huge jaw musculature could be further accommodated on the sides of their heads.

The earliest known robust ape-man species, *Australopithecus aethiopicus*, appears in the fossil record of East Africa around 2.7 million years ago. That emergence concurs broadly with the first appearance of stone tools in the archaeological record (2.6 million years ago) and makes *Australopithecus aethiopicus* a near-contemporary of *Australopithecus garhi* (2.5 million years old), the species that might have soon after given rise to the genus *Homo*. The coincidence of these three events—the rise of the robust australopithecines, the rise of incipient *Homo* (*Australopithecus garhi*), and the invention of stone tools—is exciting and complicating for paleoanthropologists. But that was not always the case.

Conventional scientific wisdom long held that Africa, between 3 and 2.5 million years ago, experienced an extreme environmental change from an abundance of forest habitats to more open, savanna habitats, and that this change caused the bifurcation of the hominin lineage into the robust ape-men and proto- or early *Homo*. The "savanna hypothesis"—the idea that adapting to open grasslands was the singular factor that drove human evolution—has always been part of paleoanthropological orthodoxy. As early as 1871, Charles Darwin argued that a major environmental change modified African ecology to the extent that some apes, our prehuman ancestors, descended from the trees and eventually became bipedal hominins. Nearly sixty-five years later, Louis Leakey, the grand old man of East African paleoanthropology, pointed more specifically to savanna habitats as the crucible of human evolution. It was a couple of decades later that Raymond Dart took up the idea that forest fruits, the preferred foods of our ape ancestors, would no longer have been available to protohominins among the grassland flora that now surrounded them—and that adaptations to exploit new edible resources and to survive in the open led to the emergence of our biological family. As the hominin fossil record filled in throughout the twentieth century, and scientists concurrently gained a more sophisticated understanding of the evolution of ancient African environments, the "savanna hypothesis" also became increasingly developed.

By the late 1950s, South African paleoanthropologist John Robinson proposed a refinement of the "savanna hypothesis" sometimes referred to as the "dietary hypothesis." At that time, some researchers assumed that Dart's *Australopithecus africanus* was a direct ancestor of the genus *Homo*, and that *Australopithecus africanus* was also contemporaneous with a robust ape-man from South Africa now called *Australopithecus robustus*. Robinson argued that *Australopithecus robustus* adapted to the emergent African savanna by maintaining a strictly herbivorous diet but diverged from its forest ancestors in specializing on the coarse vegetable matter common in open country. Such resources, like nuts and plant roots, are either extremely hard or notoriously tough to chew, and it was in this light that the large jaws and thickly enameled teeth of the robust australopithecines made adaptive sense.

In contrast, the jaws and teeth of the presumptive *Homo* ancestor, *Australopithecus africanus,* are more gracile. Robinson contended that this more lightly constructed ape-man (and, later, *Homo*) solved the problem of savanna survival by increasingly incorporating meat into its diet. Meat on the hoof—in the form of large herds of grazing ungulates—is more readily available on the African savanna than it is in forests. Meat is also soft and does not require that its consumer have a massive dental battery in order to break it down in the mouth. Meat does, however, adhere to bones, comes in large packages, and is stubbornly encased in hairy, elastic hides, so a cutting technology would have been most useful for a blunt-toothed hominin that had begun to exploit this resource. Thus, the invention of stone tools became completely explicable and assignable to proto-*Homo* with the interspecific resource partitioning between gracile- and robust-jawed hominins envisaged in Robinson's "dietary hypothesis." Further, acquiring meat presumably requires a smarter brain than does picking stationary nuts or grubbing for fixed roots. Under this construct, it was therefore not surprising that our large-brained, small-faced lineage persists today as us, and that the small-brained, large-faced robust ape-men eventually went extinct.

Of course, this model was too simple to ever hold up. Three avenues of recent investigation argue against a specialized, exclusively nut or root diet for the robust australopithecines. First, nutritional analyses of these rather low-quality savanna plant foods suggest that it would be exceedingly difficult, if not impossible, for a large-bodied primate like *Australopithecus robustus* to meet its daily caloric and protein requirements by relying solely on them. Second, incidences of microscopic scratches and pits on the chewing surfaces of robust australopithecine teeth, when compared to the damage on the teeth of modern primates *whose diets are known,* demonstrate that the South African ape-men ate many different types of food. Third, isotopic analyses of *Australopithecus robustus* teeth indicate that it was an eclectic feeder.

All plants take up two types of stable carbon (C) isotopes, $\delta^{12}C$ and $\delta^{13}C$, during photosynthesis, the process by which they capture energy from the sun. However, tropical grasses and some kinds of sedges are able to convert $\delta^{13}C$ into sugars and body tissues more easily than can tropical trees and shrubs. As a result, trees and shrubs tend to "select against" $\delta^{13}C$ during photosynthesis and thus have lower levels of this isotope in their tissues than do tropical grasses and some kinds of sedges. Thus, with regard to their proportions of $\delta^{12}C$ to $\delta^{13}C$, trees and shrubs form a distinct group as opposed to grasses and sedges. The tissues of animals that eat plants contain the same proportion of $\delta^{12}C$ to $\delta^{13}C$ as the types of plants that they consume—trees and shrubs (browse) versus grasses and sedges (graze). The tissues of meat-eating animals reflect, in turn, the $\delta^{12}C/\delta^{13}C$ ratios of the types of herbivores—browsers, grazers, or mixed browse/graze feeders—that they kill and eat.

In a groundbreaking 1999 study, paleoanthropologists Matt Sponheimer and Julia Lee-Thorp demonstrated that *Australopithecus africanus* teeth from Makapansgat, a South African cave, have $\delta^{12}C/\delta^{13}C$ ratios similar to those of meat-eating hyena teeth recovered from the same site. This suggested to Sponheimer and Lee-Thorp that *Australopithecus africanus* was sometimes carnivorous, feeding at least occasionally on the carcasses of grass-eating ungulates. More recently, a group of Italian and South African researchers diagnosed lesions on the vertebrae of a partial *Australopithecus africanus* skeleton from Sterkfontein Cave as possible traces of brucellosis, an infectious disease caused by bacteria of the genus *Brucella*. Pregnant animals infected with *Brucella* bacteria usually abort, and the infection can spread to others through contact with the fetal membranes or other bodily discharges that attended the miscarriage, as well as through ingestion of the infected animal's milk. Given this context, the study's authors venture that "it seems reasonable to suggest that some species of *Brucella* . . . could have been the infective agent in this *Australopithecus africanus* individual, through contact with (or consumption of) infected tissues of other mammals, such as parturient discharges, fetal membranes or meat of young antelopes or other Ungulata."

Skeptics of ape-man meat eating have not yet addressed the brucellosis story, but that does not mean that paleoanthropology's antique hypotheses are not still remarkably abiding. They, in fact, continue to moor some researchers, including those who argue that the $\delta^{12}C/\delta^{13}C$ ratio in *Australopithecus africanus* can be explained by other factors, such as the ape-men consuming large amounts of grass-eating insects. When a similar $\delta^{12}C/\delta^{13}C$ ratio was identified in *Australopithecus robustus*, the supposed specialized nut- or root-feeding vegetarian, appeal to non-ungulate sources of the grass-signal became even more crucial for some doubters. Researchers, working within the frame of

Robinson's "dietary hypothesis," originally classified artifacts made from antelope bones and found at *Australopithecus robustus* sites, as root-digging tools. With the new isotope results indicating a grass-eating signal, many easily shifted allegiance to a new hypothesis that the bone tools were instead used by *Australopithecus robustus* to open the mounds of grass-eating termites, which the hominins then consumed.

It was these types of intellectual gymnastics that partisans conducted to deny the obvious for supposedly primitive hominins. Under their theoretical construct (and constraints), eating the flesh and marrow of large vertebrates simply could not be attributed to the ape-men; hominin meat eating had to stay the exclusive capacity of the *Homo* lineage (*Australopithecus garhi* probably included). Carnivory is, after all, one of the primary factors that made us human—supplying the high energy and quality nutrients to grow big, smart brains, ensuring the survival and eventual primacy of our species. It seemed obvious that *Australopithecus* was a small-brained idiot that went extinct.

## SURVIVAL MODE

However, there were other researchers who were willing to consider the isotopic data more straightforwardly. If the accumulated evidence was saying that the robust ape-men probably consumed the flesh of large vertebrates, then maybe they did. But, how then could one explain the extraordinary jaws and teeth of those remarkable hominins? To answer the question requires a turn to the modern world, in which wild animals experience fat and lean times throughout a year.

Good times of year provide an abundance of an animal's preferred foods. Harsher periods challenge the animal to subsist on other, non-preferred fallback foods. Among modern apes, gorillas and chimpanzees both prefer fruit, but when seasons change and fruit becomes scarce, each can also subsist on lower-quality leaves and piths as fallback foods. However, the difference between gorillas and chimpanzees is that gorillas are able to subsist *entirely* on fallback foods, while chimpanzees cannot; chimpanzees still require some fruit to survive and function properly. Researchers Greg Laden and Richard Wrangham suggest that the extreme morphology of robust australopithecine skulls and teeth probably indicates that these hominins followed a gorilla-like strategy of relying completely on low-quality vegetable foods as fallbacks when preferred foods like fruit and meat were in short supply. Adaptations for exploiting these types of nonpreferred, low-quality resources are the ones that should be prominently expressed in the anatomy of an animal that, in lean times of the year, adopts a gorilla-like diet and foraging strategy. This is because efficient exploitation of a single (or small range of) nonpreferred food(s) is *essential* for an individual's survival during those critical time periods.

On the ancient African savanna, likely australopithecine fallback foods included the underground storage organs of plants. Following this suggestion, paleoanthropologist Nathaniel Dominy and his colleagues investigated physical properties—especially relative hardness, brittleness, and toughness—of the roots, bulbs, and corms of nearly one hundred edible African plants and matched these qualities to the chewing adaptations in skulls and teeth of various ape-man species. Their results indicated that robust australopithecines could have relied on certain hard but brittle corms, as well as on hard/tough tubers, while more ancient, less-specialized ape-man species, like *Australopithecus africanus,* may have turned to soft/tough plant bulbs as fallback foods. Not insignificantly, Dominy's research team also stressed that that many of the corms and tubers in these categories have elevated, "grass-like" $\delta^{13}C$ enrichment—perhaps paving a way for deniers of ape-man meat eating to skirt the "$\delta^{12}C/\delta^{13}C$ *Australopithecus* conundrum."

Seeds and nuts, some of which have very similar mechanical properties, were other potential fallback foods available to hominins on the African savanna. But the paucity of complex microscopic wear damage on the chewing surfaces of many ape-man molar teeth seemed to argue against small seeds and nuts as significant dietary resources. However, an analytical shift in focus—moving forward in the mouth, to the premolars—presents the intriguing possibility that larger seeds and nuts, with diameters ranging from 10 to 50 millimeters, may have served *Australopithecus* as common fallback foods. The microscopic wear on ape-man premolars has not been seriously studied, but that deficiency may not mean much.

Recently, a sophisticated analysis of the material properties of mammalian tooth enamel indicates that microscopic wear is created mostly by truly minuscule abrasives—specifically, by the grit that adheres to ingested foods and by hard but very tiny food particles that are less than 5 millimeters in diameter. A new biomechanical model of the cranium of *Australopithecus* concludes that it

> evolved in part to withstand [chewing] loads positioned specifically on the premolar. . . . [It] is not well designed for processing tough, displacement-limited foods [fibrous, squashy things like tubers] but instead represents an

adaptation to eating stress-limited food items such as seeds or nuts, in which a soft, nutritious core is mechanically protected by a hard outer casing. The casing is fractured by premolars (thereby explaining why buttressing of the face above the premolars is so important) and then is spat out, leaving the soft core to be masticated by the molars. The size of mechanically protected seeds, measured as an effective radius, typically ranges from 1–100 [millimeters].

Following this line of argument, premolars are the most important teeth to understand, because gape at the molars of ape-men—further back in the open mouth than at their wider premolar gape—was too small to handle the initial cracking of these large food objects. When ape-men used their premolars to bite into hard foods, it probably only infrequently resulted in microscopic wear on the chewing surfaces of those teeth (see previously). More often, that kind of biting likely induced stress fractures at the junction between the teeth's enamel and underlying dentine. Further, significant chips observed in the enamel of some australopithecine teeth confirm that ape-men, at least occasionally, consumed large, hard foods, like seeds and nuts.

These exciting new investigations of potential australopithecine fallback foods are obviously at some odds with each other, but a general consensus is emerging that most ape-men were flexible, broadly adapted foragers—far from the Robinsonean stereotype of *Australopithecus* as essentially a bipedal cow. For instance, summarizing their study of *Australopithecus robustus* isotopes, Matt Sponheimer and his colleagues concluded that the species "had an extremely flexible diet, which may indicate that its [specialized] masticatory morphology *signals an increase,* rather than a decrease, in its potential foods" (my emphasis). In short, the gross anatomy of ape-men skulls and teeth shows us what these animals were *capable* of eating, while the microscopic wear (and/or chips) on their teeth and their dental isotopic signatures are direct indications of what they *actually* ate. And, study after study of tooth wear and isotopes is now coming up with significant mismatches between the extreme chewing potential of *Australopithecus* expressed in its morphology and the more general nature of its diet in reality.

A challenge to this new perspective is the analysis of more than twenty isotopic samples from the 1.9–1.4 million-year-old teeth of the East African robust ape-man species, *Australopithecus boisei*. This recent study concludes that *Australopithecus boisei* was unique among the other species of its genus in having had a diet that "included more [grass/sedge] biomass than any other hominin studied to date." The grass-eating carbon isotope signal is so elevated in *Australopithecus boisei* that it is statistically indistinguishable from those of pure grazers, like horses, some pigs, and hippopotamuses. In light of this remarkable finding, the study's authors seem to reject the possibility that the grass-eating signal of *Australopithecus boisei* was the result of it having fed on grazing prey animals. Thus, unlike the rest of the ape-men species studied isotopically, *Australopithecus boisei* might have been an extreme dietary specialist who would have competed with grass-eating ungulates and monkeys for food, rather than with generalist feeders, such as other species of ape-men. (A specialized sedge diet for *Australopithecus boisei* was largely ruled out because sedges "often utilize $C_3$ photosynthesis [that is, tree- and shrub-like], are not widely distributed in many habitats, and might have been of dubious nutritional value without cooking.") Wear on the teeth of *Australopithecus boisei*—whose enamel is often so deeply worn that swaths of the underlying dentine are exposed at the chewing surfaces of teeth—is consistent with this hypothesis. This kind of extreme wear is typical damage associated with the habitual, repetitive chewing of highly abrasive foods, like grass. It is only data on *Australopithecus boisei* tooth chipping that complicate the picture; paleoanthropologist Paul Constantino and his colleagues have shown that some teeth of *Australopithecus boisei* are significantly chipped, which usually indicates powerful chomping of large, hard foods.

When the lingering incongruities over the diet of *Australopithecus boisei* are combined with previously discussed data on the rest of the genus, it directs us to abandon the archaic notion of ape-men as simply some kind of regressed, masticating automaton. That said, the reconstructed survival strategy of the "generic" ape-man *is* set off from that inferred for their contemporary hominin cousins, the earliest members of our own genus, *Homo*. Viewed through the lens of a fallback food model, early *Homo* might, in contrast to robust australopithecines, be seen as adopting a more chimpanzee-like strategy in times when there were shortfalls of preferred foods. Primates like this—who in lean times use fallback foods only to supplement, *and not completely replace,* preferred resources—are released from the pressure to specialize morphologically. Hence, the *Homo* lineage, that hominin group that employed a more chimpanzee-like strategy, shows fewer extreme specializations in its skull and dental anatomy than are seen in the ape-men, especially the "gorilla-like" robust ape-men.

We must, of course, also consider the role of emergent stone technology in such a scenario, with its potential to further release *Homo* from external evolutionary pressures. There is a complication, though: fossils of robust australopithecines are found in archaeological association with stone tools almost as frequently as are those of early *Homo*. Further, the isotopic data discussed previously seem to strongly implicate some ape-men in meat eating (at least to some degree)—and, as is argued later, stone tools were probably invented to process animal carcasses. So, the circle closes, and the question must be asked again whether *Australopithecus* could have been much more heavily involved in meat eating and predation than the dogma of Robinson's "dietary hypothesis" has allowed? Of course, we need not listen only to contemporary voices, backed by their latest high-tech approaches, for examples of bona fide opposition to the "dietary hypothesis." From the early 1950s onward, Raymond Dart was convinced of the hunting prowess of the ape-men. The "dietary hypothesis" be damned.

## MAN, NASTY AND BRUTISH

Prior to his development of the "killer ape hypothesis," Dart was caught in a nearly twenty-year impasse with the majority of his colleagues. The stalemate was over the validity of the Taung Child's purported hominin status. The prevailing notion in the 1920s was that Europe, or possibly Asia, was the birthplace of humankind. The numerous Neandertal and Cro-Magnon remains recovered in Germany and France from the mid-1800s onward seemed to confirm this idea. By 1912, a skull with a large brain capacity and an apelike lower jaw was found at Piltdown, England, meeting precisely the geographic and morphological expectations of what and where the first human should be.

Almost immediately upon its announcement, the authenticity of Piltdown Man was challenged by some in the scientific community. But, the (retrospectively) transparent chauvinism of British intellectual luminaries, such as Sirs Arthur Keith, Arthur Smith Woodward, and Grafton Elliot Smith, blinded them to what should have been an obvious forgery: the lower jawbone of a modern orangutan with crudely filed down teeth and the cranium of a medieval human—all chemically stained to simulate the patina of great antiquity—were planted together in a gravel pit at Piltdown. Sadly, the combined gravitas of experts like Keith, Woodward, and Smith drowned out the initial rumblings of dubiousness, with embarrassing and long-term consequences for the science of paleoanthropology, not the least of which was Piltdown Man's prominent role in quashing the legitimate import of

*Australopithecus africanus.* It was only in 1953, after a trio of resolute dissenters, Wilfrid Edward Le Gros Clark, Kenneth Page Oakley, and Joseph Weiner, subjected Piltdown Man to microscopic and chemical analyses—which confirmed its hodgepodge, felonious origins—that the "fossil" was undeniably proven as and accepted to be a hoax.

However, Dart and the Taung Child were not yet out of the woods. Since the announcement of *Australopithecus africanus,* Dart had always contended with more than just Piltdown Man's impediment to the academy's acceptance of the Taung Child. Nearly every aspect of Dart's report on the find was condemned in the halls of the English establishment—from his writing style (to be sure, it is wincingly ornate in some cases) to his etymological gaffe of combining Greek and Latin roots in constructing *Australopithecus africanus* as the scientific name for his newly discovered species. After Piltdown Man's fall, many of these critiques were placed aside as the pettiness they, in fact, were—but, even still, Dart could not shake the incessant, wagging reprimand of one Lord Solly Zuckerman. South African by birth, Zuckerman became a British national and more, eventually receiving a knighthood in 1964 and then entering the Peerage of the United Kingdom in 1971, with the title of Baron Zuckerman of Burnham Thorpe. Zuckerman paid his noblesse oblige prospectively, serving the British government first as Scientific Advisor to the Ministry of Defence from 1960 until 1966, and also as Chief Scientific Advisor of the entire government between 1964 and 1971. Like Dart, Zuckerman was horrified by man's penchant for unchecked aggression, but unlike Dart, Zuckerman's published views on the topic often deviated from the academic into quasi-activism; the titles of Zuckerman's two-volume memoirs—*From Apes to Warlords,* and *Monkeys, Men and Missiles*—reflect his distain of war and his very public support of nuclear disarmament.

Zuckerman also differed from Dart in failing to discern that the humanness of *Australopithecus* is nuanced anatomically, as *should* be expected in such a transitional animal. Cloaked in a guise of quantitative rigor, Zuckerman provided detailed measurements comparing *Australopithecus* remains to the bones of modern apes and humans, presuming that there existed some kind of exact morphometric Rubicon separating pongid and hominin. His statistical assailment against *Australopithecus* began in 1928 and persisted into the 1950s, well after the mainstream of paleoanthropology had welcomed australopithecines into the family of man. As a result, Dart was unusually vocal against a critic in expressing often and unequivocally his abhorrence of the tenacious Zuckerman.

Except for his seething opinion of Zuckerman (maintained until the end of his long life in 1988), Dart was otherwise largely unresponsive to the boorish treatment the Taung Child received from his colleagues. And, although fossil recovery from South African cave sites continued unabated from the 1920s onward, the work was conducted by researchers other than Dart. In particular, the 1930s and 1940s belonged to a flamboyant personality, next to which even Dart paled in comparison. The grandfatherly facade of Robert Broom belied a frat-boy deportment, when, at age sixty-eight, he entered the field of paleoanthropology after previously questing the world's earliest mammal fossils. Supporting himself as an itinerant doctor, Broom scoured fossil beds in the Australian outback and then in the empty Karoo region of central and western South Africa, all the while cutting an alternatively dashing and shocking figure in the vast emptiness of those places. On some occasions he excavated in full formal attire, down to the wing-collared shirt; on others he took advantage of what he considered the health-giving properties of the sun and worked completely unclothed. And although Broom's idiosyncrasies ran infamously beyond nudity (even as a geriatric he incensed fathers and affronted husbands on his womanizing run through podunk South Africa), it was still the fossils to which he returned with his greatest adulation. In testament to that love and to the merits of tenacity, Broom completed his monographic treatment of *Australopithecus robustus* (then called *Paranthropus crassidens*) on his deathbed. The task sealed, he relinquished tersely (and, by best accounts, *not* apocryphally), "Now that's finished . . . and so am I." That dying witticism closed a life that, in retrospect, is defined by Broom's defense of Dart, of the truth of Darwinian evolution, and of the African origins of man.

Dart's generally measured reaction (and sometimes non-reaction) to the establishment's rejection of the Taung Child does not mean that he was completely unfazed. Indeed, by 1943 personal stresses so exacerbated the lingering inanities of his colleagues that he suffered a mental collapse. And, although he remained productive throughout the 1930s and early 1940s in his capacities as a professor and neuroanatomist at Wits, it is clear that his enthusiasm for human paleontology waned significantly during this period. Indeed, the field was never Dart's infatuation, even after the Taung Child fell into his lap. It might also have been that upon meeting Robert Broom, Dart felt that the torch had been passed to a more fully invested defender of *Australopithecus africanus*. Finding an adult of the Taung Child's species became Broom's stated goal, as he was well aware that without an adult fossil of the species, *Australopithecus africanus* would be confined in taxonomic limbo (anatomists have long recognized that juvenile apes and juvenile humans are more similar morphologically than are adults of both animals—who, once more fully grown, develop, respectively, distinctively ape or distinctively human anatomy; thus, sorting taxonomically unknown juvenile remains into species is less clear-cut than is assigning fossils of adults). Broom had the paleontological chops, honed in some of the remotest country of Australia and South Africa, to suggest that if anyone could attain such a goal, it would be him. But, it is just as likely that Dart's tacit approval of Broom may have occurred on a much more visceral level. Dart, himself, was certainly not a staid man (indeed, it must have been a giddy student body that attended Dart's classes, hopeful each meeting that he would once again break from his lecturing into some crazed display of athleticism: in a favorite sequence, Dart sprung off his podium, seized the lecture hall's overhead water pipes, and then swung across the stage—much better to *demonstrate* rather than drone about the rotary capabilities of the human shoulder), but the admiring shock of meeting Broom is evident in his recollection: "Broom immediately wrote a letter of congratulations [in response to the announcement of the Taung Child,] and two weeks later burst into my laboratory unannounced. Ignoring my staff, and me, he strode over to the bench on which the skull reposed and dropped on his knees 'in adoration of our ancestor' as he put it. He stayed with us over the weekend and spent almost the entire time studying the skull. Having satisfied himself that my claims were correct, he never wavered."

Broom's search for an adult *Australopithecus africanus* was not without drama. His innate irascibility and disdain for authority had long before conspired against him, and by the early 1930s he was unemployed and destitute. Unbendingly principled on matters of science, and ever true to the stereotype of the recalcitrant Scot, Broom had refused to drop his instruction in evolution and therefore was summarily ejected from his professorship at the University of Stellenbosch in 1910. Broom never found another job at a South African university; all were government run and religiously connected. His subsequent peripatetic existence must have been grinding on a man in his late sixties—even one as energetic as Broom—when finally in 1933, through the intervention of Dart, he received a scientific post at the Transvaal Museum in Pretoria.

It was from there that Broom was at last able to launch a serious search for additional *Australopithecus africanus* fossils. An adult cranium was finally recovered from Sterkfontein Cave in 1936. Broom's analysis of it confirmed Dart's contention that *Australopithecus africanus* was, in fact, a hominin. Other remarkable finds followed in rapid succession, with Broom proposing three newly discovered species of early hominins from Sterkfontein, as well as from the nearby cave sites of Kromdraai and Swartkrans. (Most experts now consider that Broom's finds belong to just two distinct species, *Australopithecus africanus* and *Australopithecus robustus*. In 1949, Broom's younger colleague, John Robinson [of the "dietary hypothesis" fame], identified from Swartkrans the first known early South African fossil of the genus *Homo*.)

In 1945 Dart was finally pulled back into the fossil game by an eager Wits pre-med student named Phillip Tobias, who would eventually go on to his own considerable repute. Until his death in 2012, at the age of eighty-six, Tobias was considered the dean of South African paleoanthropology, always called upon to proffer his opinion and dominating local news coverage of any significant new find. But, his first noteworthy contribution to the field was his leadership of an all-student expedition to the distant cave site of Makapansgat, in South Africa's remote Northern Transvaal. The paleontological outing recovered a haul of important monkey fossils and then reported back to Dart, Tobias's professor. Under Dart's auspices, Makapansgat was soon yielding *Australopithecus africanus* fossils.

Most interesting to Dart, though, was the large sample of non-hominin bones also recovered at Makapansgat. Dart discerned clear patterns in the Makapansgat fossil assemblage: only select bones of dozens of antelope skeletons were preserved, and of those bones, only certain parts of them remained. Dart's interpretation of these fossil patterns was that *Australopithecus africanus* had killed the animals from which the bones derived, consumed the animals, and then selected choice parts of their skeletons to use as tools and weapons. Dart's voluminous writings about these alleged implements reveal the audacity of his imagination, in which great assumptive leaps replaced careful inference: the base of an antelope horn, for instance, fit in Dart's hand as comfortably as did the hilt of a knife, so antelope horns were obviously ape-man stabbing tools; antelope leg bones have heavy, knobby ends that were, to Dart, quite clearly used as clubs; and so forth. Dart believed that *Australopithecus africanus* accumulated these "osteodontokeratic" (bone, tooth, and horn) tools and weapons in Makapansgat during its occupation of caves. The living presence of ape-men at Makapansgat was supposedly further verified by black-colored bones that Dart interpreted as having been charred in carefully tended campfires. Dart thus imbued *Australopithecus africanus*—a protohuman species possessing a brain only slightly larger than that of a chimpanzee, and which we now know lived about 3 to 2 million years ago—with astonishingly humanlike behavior.

Even more staggering was Dart's analysis of the *Australopithecus africanus* and monkey fossils from Makapansgat, as well as those from the site of Taung. He noted that these primates were represented almost exclusively by skulls and that many of those skulls have massive crushing damage to their sides. The explanation for all this was obvious to Dart, and he created a picture of *Australopithecus africanus* as vicious cannibals, pulverizing the heads of their cohorts with clubs that they had fashioned from antelope leg bones. Not content with simple slaughter and the consumption of its own kind, the Makapansgat ape-man reveled in its murderous sport like some kind of prehistoric serial killer—collecting its victims' skulls in the cave as grisly trophies. *Australopithecus africanus* had become the original "killer ape."

In Dart's time, such a fanciful reconstruction of violent behavior in early hominins was beyond the ken of even the most informed of the academic community. Modern scientific study of living primates was in its infancy in the late 1940s, and the scant data that *were* available suggested that apes were essentially placid vegetarians. This was in marked contrast to modern humans, who engage in all manner of heinous violence, from garden-variety homicide to full-scale warfare. To see traces of these unfortunate human habits in the fossil residues of the then–earliest known protohuman species, presumably just emerged from the apes, was completely unexpected. But, that is *exactly* what Dart contended. At the very inception of our lineage we exceeded every other animal in all manner of unrestrained mayhem and brutality. If Dart was right, it was, ironically, cruel inhumanity that was the very underpinning of our humanness.

And, viewing the world through Dart's eyes, it was all too apparent that modern man retained more than just mere vestiges of our "blood-besplattered, slaughtergutted" past. The "killer ape hypothesis" germinated amid two great world wars that arose in the gangrenous agony of trench warfare and only ceased beneath the utter annihilation of a mushroom cloud. Between those ghastly markers, people were cooked in ovens, gassed to death, and experimented upon as if they were lab rats. And it wasn't just Nazi atrocities to which mid-twentieth-century people were victims, participants, accomplices, or simply witnesses. Concurrently, aborigines in Dart's native Australia were still denied citizenship and treated as chattel. In South Africa, Dart's adoptive country, de facto slavery was being institutionalized and legalized under an apartheid regime that was organizing out of centuries of white oppression and sadism. The evolutionary soul of humanity seemed a moonscape, barren of any inherent morality. Abandoning all constraint imposed by scholarly convention, Dart was typically lucid *and lurid* in making a précis with the "killer ape hypothesis": "On this thesis man's predecessors differed from living apes in being confirmed killers: carnivorous creatures, that seized living quarries by violence, battered them to death, tore apart their broken bodies, dismembered them limb from limb, slaking their ravenous thirst with the hot blood of victims and greedily devouring livid writhing flesh." Ironically, it was Robert Ardrey, an American dramatist (and Dart's mouthpiece in four popular books), who provided the voice closest to cool detachment when he abstracted the "killer ape hypothesis" thusly: "Man is a predator whose natural instinct is to kill with a weapon." In no subtle way, predation and aggression were coupled as the ultimate propellants of human evolution.

# Tamping the Simian Urge

. . . but still there was about him a suggestion of lurking
ferocity, as though the Wild still lingered in him and the wolf
in him merely slept.

—Jack London, *White Fang*


*Imperare sibi maximum imperium est.*
(To rule one's self is the ultimate power.)

—Seneca the Younger


The "killer ape hypothesis," as construed by Dart and Ardrey, had
profound worldwide impact. But, as anthropologist Matt Cartmill
carefully documented, the idea was not without precedent. Cartmill
traced the notion of protohuman killer apes all the way back to at
least the late 1880s. The American writer Charles Morris concluded
that hunting required a wilier, and perhaps even devious, primate.
Even more explicit than Morris in positing causal linkage between
human hunting and our species's general truculence were Harry Campbell and Carveth Read, who each, independently, published several
articles and books on the topic between 1904 and 1925. The same
nauseating waves of cannibalism, unquenchable bloodthirst, cruel
misogyny (specifically), and raging misanthropy (generally) that
course through the writings of Dart and Ardrey also typify the pre-
Dartian ramblings of Morris, Campbell, and Read. However, these
earlier voices that lobbied for a fundamental human barbarity are
today all but unknown. Instead, the better-appreciated substantiation
for our emergence from a killer ape parentage was forwarded by Dart's
contemporary, the Austrian ethologist Konrad Lorenz, who was infamously resigned to his conviction that humans are essentially deviant,
maladapted assassins.

It seems that Lorenz came by that view honestly. His lifelong connections to human aggression and human cruelty were complex. There

is no denying, as some have tried, that Lorenz's writings served the Nazi propaganda machine, including his scientifically masked support of Hitler's racial hygiene policies. And, Lorenz was a card-carrying member of the National Socialist German Workers' Party in the late 1930s. Eventually, after World War II, he did denounce the Nazis and apologized publically for his tacit collusion, but he still dissatisfied many of his critics in never fully condemning interpersonal violence and institutional brutality (in *Das Sogenannte Böse: Zur Naturgeschichte der Agression* [*On Aggression*], his definitive 1963 statement on how various forms of animal aggression are linked, Lorenz wrote, "I believe that present day civilized man suffers from insufficient discharge of his aggressive drive"), and in never fully abandoning views that teetered toward the eugenic (including, in his late-life book *Civilized Man's Eight Deadly Sins,* the admonishment of imminent human genetic decline).

Slightly less controversially, Lorenz is also associated with the behavioral concept of imprinting (that is, the untaught, seemingly automatic process by which a young animal recognizes and is attracted to another animal, usually its mother)—so much so that one of the most recognizable images of him does not even show his face, but instead the stooped gait of an aged man with his back to the camera, leading a trail of imprinted greylag geese along a wooded path. Imprinting, as a type of instinct, was central to Lorenz's thinking that much of animal behavior is instinctive. In Lorenz's view, instincts, as spontaneous manifestations, demand outlet for the proper functioning of an animal (hence the previous quote from *On Aggression*). Consistently deny the organism that vent, and systemic psychological and physiological failure surely looms. Add to this that Lorenz's survey of aggression across a broad array of vertebrates convinced him that aggression was merely another widespread instinct in the animal kingdom, and the quote from *On Aggression* takes on an even more chilling connotation. Killer apes should be expected to kill; it is natural. Moreover, killer apes *need* to kill; it is natural. Deny this need and humans *and human society* become sick.

While plenty of specialists have questioned Lorenz's basic assertion of the aggression instinct, it was his higher-order inference—that the civilized human condition is fundamentally one of severe repression—that caused the most stir in the broader circle of the scientific laity. And Lorenz's message got even worse in the fine print, where he proposed further that dehumanization of nongroup members by people leads to a unique kind of intraspecies aggression in humans, culminating ultimately in murder and war. In effect, Lorenz was claiming that human killing *was* predation because victims were not seen as conspecifics by the killers but were rather deemed, at least subconsciously, a separate species (borrowing a phrase from the developmental psychologist Erik Erikson, Lorenz referred to this phenomenon as "pseudo-speciation"). In other words, humans possess an odd form of aggression played out within the species but with other characteristics of the type of aggression employed by some animals in hunting.

Two aspects of this contention are interesting. First, it seems to contradict Lorenz's own extended argument in *On Aggression* that predatory aggression and social aggression are dependent on distinct motivations. Second, it anticipated some important aspects of primatologist Richard Wrangham's ideas about lethal raiding by "demonic male" humans and chimpanzees, in which all manner of aggression is seriously conflated. The entanglement is well illustrated in Wrangham's contrasting of chimpanzee and bonobo behavior. Chimpanzees (*Pan troglodytes*), bonobos (or, pygmy chimpanzees as they are sometimes called) (*Pan paniscus*), and humans shared a most recent common ancestor some 8 to 4 million years ago, while chimpanzees and bonobos shared a much more recent common ancestor, perhaps only one million years ago. However, bonobo sociality contrasts markedly with that of chimpanzees. First, females are dominant in bonobo groups, establishing and maintaining their dominance over the slightly larger males through all-female coalitions. Second, coalitional bonds are cemented by female–female food sharing and face-to-face sexual contact, during which they rub their genitals together. And, in general, bonobos engage in sexual contact frequently and casually, in many cases to seemingly allay what could otherwise turn into aggressive interaction. Indeed, serious—and especially lethal—aggression is virtually unknown in bonobos. Wild bonobos also hunt less frequently than do wild chimpanzees. Wrangham asserts a causal link between the lack of social violence and the low frequency of predation by bonobos, just as he, in contrast, links functionally the predominance of social violence to the high frequency of predation by chimpanzees—going as far to intimate that bonobos might somehow be more "sympathetic to a victim [of predation]." That's quite a declaration, but even more profound is the tacit supposition that a predatory primate is also necessarily an inherently violent primate.

## SCIENCE'S COVER GIRL AND HER DARK-HEARTED WARDS

It was, of course, Jane Goodall who first witnessed the nature and extent of chimpanzee brutality. Toward the goal of better understanding the extant apes and thus (he hoped) elucidating our own origins and uniqueness, Goodall was recruited in 1960 by Louis Leakey to be one of his three "Angels," a cadre of young women that also included Dian Fossey to study wild gorillas and Biruté Galdikas to study wild orangutans. The photogenic Goodall was an instant media darling, the wholesomely inviting face of *National Geographic*. Even more impressive was her intrepidness: Who before, much less a pretty young Englishwoman, had ever committed so completely to studying the behavior of animals as to live among a group of wild apes at a place like far-flung Gombe, a remote Tanzanian field site on the northeastern shore of Lake Tanganyika? That dedication, combined with a voracious curiosity, paid off immediately for Goodall. In her initial research season, she became the first scientist to ever observe chimpanzees modifying and using twigs as tools to "fish" edible termites out of their earthen mounds. Leakey's response to that discovery indicates its profundity: "Now we must redefine tool, redefine man, or accept chimpanzees as humans." It was widely assumed prior to Goodall's observations that tool manufacture and tool use were the exclusive domain of humanity.

And, termite foraging was not the only surprising habit of chimpanzees that Goodall was the first primatologist to report. The second of her unexpected early findings, which she announced in 1962, was that chimpanzees also hunt and eat large vertebrates, like bushpigs and monkeys. If Leakey was right about the taxonomic and philosophical implications of chimpanzee tool use, then Goodall's recognition of chimpanzee predation served only to further erode the presumed boundary between man and ape. But, even this revelation was mere foreshadowing to what Goodall announced out of Gombe and to the world in 1974—an announcement that seemed to finally and completely annihilate the duality of "us (human) and them (ape)." It was at this time that Goodall was forced to acknowledge the beginning of a four-year gang killing spree at Gombe, with members of one chimpanzee community methodically seeking out and massacring every male of a smaller splinter group (figure 3). The murders were appalling in their calculation and viciousness. Because the perpetrators approached with great stealth and in silence, victims were caught unaware. Invariably, those victims were also outnumbered, their horrific shrieking ignored as a squad of testosterone-fueled attackers pummeled them into bloody pulps. The

**FIGURE 3.** The ferocity of male chimpanzees on display at Ngogo, Tanzania. (Photograph courtesy of John Mitani)

wounded were left to die in slow agony, and Gombe's image was suddenly transformed. No longer an idyllic glade, host to frolicking troglodytes, the forest had turned gothic, suffocated in a thick fog of menace. Even Goodall, the chimpanzees' greatest advocate, "struggled to come to terms with the new knowledge" of their atrocity.

If we presume chimpanzees are living fossils, faithfully representing the protohuman condition, then Goodall's discovery of systemized carnage among them bears out Dart's (and Morris's, Campbell's, Read's, and Lorenz's) prognosis that humanity's violent disposition is a deeply entrenched one, with us from the very earliest stages of our lineage; moreover, it supports the notion that an aptitude for cruelty was the single most important requisite for the incredible evolutionary success of our lineage. But, of course, chimpanzees are *not* living fossils. They are, instead, their own unique species with its own unique evolutionary history that extends back into time as far as ours, possibly 8 million years into the past, when we each split and went our separate ways from our most recent common ancestor. This means that the propensity of chimpanzees for violence needs to be understood on its own terms. Recognizing similarities in chimpanzee and human interpersonal violence, and the ways in which violence might be linked to hunting, will never resolve the "killer ape hypothesis," but it might well be useful in evaluating its plausibility.

Research by other primatologists on the very same chimpanzees studied by Goodall, as well as that on other groups of chimpanzees, has confirmed that the four-year killing spree at Gombe was not an anomaly. Intraspecies killing among chimpanzees is not only documented at other African field sites but was even witnessed in a population of captive chimpanzees at the Arnhem Zoo, in The Netherlands. In *Demonic Males,* a comparative (and highly provocative) report on primate aggression, Richard Wrangham and Dale Peterson stress a striking commonality of intraspecies aggression in humans and in chimpanzees: it does *not* always escalate from interpersonal conflict. Instead, it is often expressed through "lethal raiding" by males, in which "hunting and killing rivals . . . is akin to predation." Such lethal raids are usually carried out in situations of existing intergroup hostility and when the aggressor party possesses strength of numbers (usually at least 3-to-1 odds), assuring their success and reducing the probability of successful retaliation by the victims.

This type of aggressive lethal raiding, conducted by both chimpanzees and humans, is an oddity compared to hostile expression in most other animals. So, trying to place the behavior in biological context is not straightforward, and it also begs the question: What is the evolutionary benefit of lethal raiding to the attackers? Of course, for any sexually reproducing individual to be successful evolutionarily—to have high Darwinian fitness, contributing a disproportionately greater amount of genetic material to subsequent generations than its conspecifics—that individual must first simply survive in order that it can gain opportunities to mate and pass on its genes. That obviously means that the individual must, at a minimum, be adequately nourished and must avoid being killed. Outcompeting and, in some cases, outright eliminating intraspecies competitors—including those from nearby outside groups—help in both these matters. We can apply these principles to explain lethal raiding by chimpanzees and humans. Lethal raiding, by presumably asserting absolute dominance over a neighboring population, preemptively eliminates dangerous male competitors and provides increased and improved access to limited food and limited females. With the adult males of a neighboring population terminated, the conquerors can move into the vacated territory and absorb females of the defeated group into its own. Thus, in theory, successful lethal raiding would seem to meet the proximate requirements leading to ultimate evolutionary success.

Since conducting field research on lethal raiding by humans poses a serious ethical predicament, the Darwinian advantages of lethal raiding are much more obvious in wild chimpanzees than in people. For instance, the observations of field biologists, that coalitions of male chimpanzees work together to bar outside males from mating with their females, is borne out at some sites by paternity tests of baby chimpanzees, who are genetically related only to within-group males. Another measure of the evolutionary importance of the defense and acquisition of territory *to both sexes* is that female chimpanzees reproduce more quickly within larger territories.

In this light, lethal raiding is a completely predictable and evolutionarily "rational" behavior—no matter how repugnant it might seem on the moral face of it. As such, it is reasonable to hypothesize that lethal raiding *should* occur in primate species, including humans, in which males can get away with it. Primeval warfare would thus appear well rooted among those close ape cousins, chimpanzees and humans. By "primeval warfare" I do not mean pitched, full-scale battles, but instead predetermined routs of an outmanned opponent. We might well consider it deadly interpersonal bullying—what Richard Wrangham encapsulates as his "imbalance of power hypothesis" for ape violence. And, it *might* be the case that the evolution of large-scale interpopulational violence among humans—beyond simple ambushes, revenge killings, and the like—is linked to this smaller-scale phenomenon. Wrangham certainly contends that this is the case. Further, in the summation of his "imbalance of power hypothesis," he also rewelds the whole ugly matter of primate violence, at all levels, to predation: "[Natural] selection has favored a hunt-and-kill propensity in chimpanzees and humans, and . . . coalitional killing has a long history in the evolution of both species"; this "hunt-and-kill propensity" of chimpanzees and humans is regarded as "akin to predation." So, predation—ostensibly, simply one of many ways to acquire sustenance—is, once again, seemingly inevitably, entangled with primate brutality.

## THE TRIUMPH OF SELF-POSSESSION

As with lethal raiding, males are the main participants in chimpanzee hunting. All chimpanzees and many, if not most, modern human hunter-gatherers are characterized by male philopatry, a social system in which males stay their whole lives in their natal groups. This means that groups are frequently composed of subunits of adult brothers and half-brothers, individuals who are significantly genetically similar. This close genetic relatedness often acts to cement long-term relationships between males and to ensure collaborative actions that will enhance not just individual fitness, but also the inclusive fitness of all relatives in the subunit. This phenomenon is yet another factor underlying the coordinated pursuit of dangerous "prey" by a group of male raiders or hunters.

And, just as intergroup raiding involves a key element of excitement in its execution, so does the way in which chimpanzees *typically* hunt. Indeed, the instantaneous shift between the stalking phase of a characteristic chimpanzee raid or hunt, with its silent discipline, followed by the furious intimidation and interpersonal disregard of the attack phase, suggests a kind of *consummation* rather than simple culmination to both raiding and hunting.

Certainly, terror and dehumanization are also components of lethal raiding by humans, but human males seem to diverge from the chimpanzee pattern of transferring this kind of unchecked thrill to hunting. Yes, hunting is stimulating for humans. Across the United States, each late autumn and early winter sees innumerable deer hunters infected with buck fever, an affliction that presents symptoms that are alternately comical (potshots at hapless—and, I might add, completely antlerless!—Guernseys; self-conducted digital amputations, assisted by wayward lead) and tragic (adrenaline-induced cardiac arrest; the Dick Cheney "friendly fire" treatment). And, as paleobiologist R. Dale Guthrie emphasizes in his book *The Nature of Paleolithic Art,* modern hunter-gatherer men worldwide are preoccupied with planning and reliving hunts: "They give graphic descriptions of their most recent hunts, which remind them of incidents from past hunts, and then of tales their ancestors told. These discussions go on hour after hour and are captivating to younger boys."

But, those same experienced hunters, who hunt as an integral part of their livelihood, have a mostly deliberate approach to the act of killing prey—even if that approach might arguably be *guided* by "a thirst for the jubilation of the seek-and-kill experience." It's not hard to grasp the reasons for this carefully calculated type of hunting. Living off the land, as do traditional foragers, is an unforgiving enterprise, subject to vagaries that almost never touch the insulated lives of urbanized Western people. Real nature rarely suffers fools. Inclement weather, lurking predators, and even simple infections are all potentially mortal eventualities in the daily life of a hunter-gatherer. Likewise, the distribution, abundance, and availability of food across space and time shape a hunter's foraging decisions, and those decisions, in turn, affect survivorship for him, his family, and his social group. In that kind of system, an animal carcass—with its dense concentration of nutrients and calories, in the form of skin, meat, marrow, and brain tissue—is the highest-ranked (or second-highest-ranked, with delicious, sugary honey sometimes topping it) among a capriciously oscillating range of food choices. It is true that locating this top-ranked resource sometimes involves a good degree of providence. But, more often it is cool-headed calculation—in, for example, patiently manning a hunting blind along a predictable game trail; in taking a careful, one-chance-only shot with the bow; or, in knowing when to abandon tracking a nonfatally wounded animal in favor of a more reliable fallback plan—that yields the greatest dividends for the hunter.

For men in traditional societies, hunting success also means something beyond mere subsistence. It confers status, because as a top-ranked food, animal product is as desired by women as it is by men. But, because most hunter-gatherers practice a sexual division of labor, women are usually relegated, day in and day out, to the tedium of plant foraging; often, as in the case of many subtropical African foragers, this means having to bend over a sharpened wooden stick for two or three hours a day of back-breaking digging in order to excavate deeply buried tubers, which serve as important (and flavorless!) dietary staples. Thus consigned, most forager women must rely on male hunters to share meat and marrow with them. Obtaining nutrient-laden, energy-packed meat and marrow contributes to a woman's and her dependent offspring's good health and survivorship. Thus, a proficient hunter's high status, in turn, holds the potential to be converted into increased Darwinian fitness—but the ways in which that happens are not always as direct as male–female exchange of meat for sex. In some foraging groups, good hunters *do* enjoy higher frequencies of extramarital matings than do poorer hunters, but the complexity of modern human culture ensures that the evolutionary advantage of being a skilled hunter is more nuanced in most cases. Research among a group of northern Tanzanian hunter-gatherers, the Hadza, illustrates the point: the best Hadza hunters have younger, more fertile, and harder-working wives than do less successful hunters. In other words, a Hadza man's predatory prowess is linked to his success in attracting the best mates—those women who, in other words, will have the highest number of viable offspring and who will work hardest at ensuring the survivorship of those offspring.

Expertise in hunting the large, warily dangerous prey of human foragers and cashing in on its concomitant evolutionary rewards does not mature from the hell-bent approach employed by chimpanzees to dispatch their prey. Application of brute physicality is an efficient means for chimpanzees to kill because they hunt in groups, they concentrate on much smaller animals than themselves, and they rely on their super-human strength and agility to overpower their victims; an adult male chimpanzee is at least as twice as strong as a grown man, and chimpanzees make many of their kills on monkeys high in the forest canopy. A human has no hope of out-muscling, out-running, or out-climbing his typical prey, but, if his mind stays clear, he can absolutely count on out-thinking those animals.

Indeed, it might be that part of the long-term evolutionary trade-off for the increasing growth of the human brain through time was concession of some muscular power. Paleogeneticist John Hawks muses about intriguing new data emerging from genomics:

> Even though chimpanzees weigh less than humans, more of their mass is concentrated in their powerful arms. But a more important factor seems to be the structure of the muscles themselves. A chimpanzee's skeletal muscle has longer fibers than the human equivalent and can generate twice the work output over a wider range of motion. In the past few years, geneticists have identified the loci for some of these anatomical differences. One gene, for example, called MYH16, contributes to the development of large jaw muscles in other apes. In humans, MYH16 has been deactivated. (Puny jaws have marked our lineage for at least two million years.) Many people have also lost another muscle-related gene called ACTN3. People with two working versions of this gene are overrepresented among elite sprinters, while those with the nonworking version are overrepresented among endurance runners. Chimpanzees and all other nonhuman primates have only the working version; in other words, they're on the powerful, "sprinter" end of the spectrum.

Hawks develops the theme elsewhere: "Genetics may be starting to make the 'expensive tissue' story [see chapter 1] come down to muscle instead of gut reduction—if I'm going to make predictions, I would say that MYH16 will not long be alone as a gene corresponding to human muscle reduction."

Alan Walker, the discoverer of the Nariokotome Boy (chapter 1), tackles the issue of human wimpiness from another, related angle, hypothesizing that nonhuman apes display great muscular power because they have "many fewer small [muscle] motor units than humans, which leads them, in turn, to contract more muscle fibers earlier in any particular task." *Pow!* Instantaneous, focused power applied to the commission of an act—a system that yields good results for chimpanzees and other apes. Why wouldn't humans, their close relative, be as powerfully capable? Walker postulates that human concession of muscular strength came at the gain of muscular control. If humans really *do* have relatively more small muscle motor units than do other apes—and this is very likely—that means, in turn, that people have, collectively, "a much greater range of motor unit sizes over their muscle masses, and this allows us to recruit muscles for more complex but less forceful tasks." This type of fine control of muscular force is critical "for effective human activities such as running, throwing, and manipulation, including tool making."

Of course, as qualified earlier, all the brain power and fine motor control in the world aren't worth a damn to a human hunter if his brain's commands are overridden by emotion. Clear thinking in survival situations—and what is a hunting and gathering life if not a daily struggle for survival?—is dependent on control of emotion. With *Deep Survival* Laurence Gonzales has written a compelling book that examines why some people succeed in survival situations and others do not. He refers to "cool" as a primary contributor to success. Gonzales is not talking about what pathetically passes for cool in some suburban teenage milieu, but about *real* cool—that ability to rein in one's emotions under the most extreme and often instantaneous stress. Real cool drops a clamp on hysteria, and good traditional hunters have real cool. They are matter-of-routine survivors, living out a largely composed existence along a razor's edge. For human hunters, brains and calm appraisal are at a premium in their development as predatory experts worthy of high social repute.

## THE METHODICAL APES

Hunting weaponry—one product of human ingenuity, control, and dexterity—is another important factor in allowing people to tackle prey that are much larger in body size than is the individual hunter. With the cumulative realization of chimpanzee tool use, hunting, and interpersonal brutality—all behavioral traits that were once the defining characteristics of humanness—the hypothetical wall between human and ape has crumbled since the 1960s. Perhaps in an effort to maintain at least a semblance of behavioral distinction between "us and them," some scientists still insist on clinging to the remaining (seemingly less consequential) disparities. Hunting with weapons was one such vestige of supposed human uniqueness. But, recently primatologist Jill Pruetz saw to toppling even this minor remnant of presumed human exceptionalism. Using their teeth to sharpen the ends of sticks into points, the chimpanzees of Fongoli, in the West African country of Senegal, fashion what are essentially simple thrusting spears, the longest of which are only a couple of feet in length. The Fongoli chimpanzees poke these simple spears into hollows in trees in an effort to stab and extract bushbabies, the small nocturnal primates who sleep in the holes during the day. Of course, these relatively flimsy weapons and diminutive prey don't come close to, respectively, the sophisticated artillery and monstrous quarry of some modern human hunter-gatherers. But, a hand-captured bushbaby, about to be eaten by a chimpanzee, *will* put up a fight. And although it would also be virtually impossible for a bushbaby to outright kill its attacker, its counterattack could, all the same, pose a lethal threat to a chimpanzee. Any open wound in a natural environment, even a seemingly inconsequential bite from a bushbaby, has the potential to go septic and thus prove fatal to the wounded chimpanzee.

Chimpanzees who gang-hunt monkeys at Gombe face the same hazards as do the Fongoli chimpanzees, only multiplied because the larger monkeys that they hunt (equipped with big, sharp canine teeth) are much more formidable prey than are bushbabies. Additionally, the Gombe chimpanzees lack the advantage conveyed to the Fongoli apes by their crude thrusting spears—the advantage of at least minimal physical separation between predator and prey during the capture phase of predation. That physical separation seems to have a remarkable composing effect on the expressive state of the Fongoli chimpanzees, as compared to the raw emotionality displayed by weaponless chimpanzees, when they (like those at Gombe) are pummeling a resistant victim to death. The Fongoli chimpanzees thus, by nature of their inventiveness in creating and employing hunting weapons, behave in a very atypical way for a chimpanzee capturing and killing its prey. As a relevant aside, I have been with Hadza hunters on several occasions when they have captured and killed little bushbabies in essentially the same way as do the Fongoli chimpanzees. A Hadza man will calmly poke an untipped, but sharp-pointed, arrow into a bushbaby sleeping hole, stab and pull the poor thing out on the end of the arrow, and then crack it repeatedly over the head with another arrow until it dies. The bushbaby fits comfortably in the man's hands once dead—but, like the Fongoli chimpanzees, that man knows a bite from the little creature would, at the very least, smart.

Surprisingly, Pruetz's unforeseen revelations are, at best uncelebrated and, at worst derided by some of her primatological peers. As if it somehow negates the magnitude of her completely unexpected observations, the small number of hunts initially reported by Pruetz (since her 2007 publication on spear hunting, she continues to regularly observe the behavior at Fongoli) was a toehold for condemnation. Other critics fixate on the fact that bushbabies are small and essentially defenseless, so their capture by (predominantly female and juvenile) chimpanzees should be considered gathering rather than hunting. But, specialists who denigrate the Fongoli behavior as mere gathering muddy the essential—and wholly dissimilar—characteristics of vertebrate animals versus plants as prey. As stressed previously, unlike a plant, which is sessile and can be *truly* gathered without retaliation against its gatherer (unless, of course, it is poisonous or thorny, imbuing it with a sort of "passive retaliation"), even a puny bushbaby is likely to give some fight for its life if it is unable to flee and is unintentionally afforded the opportunity by its hunter. A chimpanzee attempting a capture by hand would provide such a counteroffensive opportunity to the bushbaby, while spearing it removes the opportunity, putting injury-reducing distance between predator and prey.

Chimpanzees also use simple stick tools to probe termite mounds and anthills. The bite from a safari ant is not life threatening, but it is quite painful. An animal that experiences pain, especially as a consequence of counterattack, is likely, out of surprise and fear, to react emotionally (read, aggressively, here). Preemptively obstruct pain (read, with space-creating weaponry, here), and a predator has a chance to maintain its composure, its cool. Perhaps this is an ethologically unorthodox way to consider primate foraging, but that should not nullify the appreciation that each edible resource presents its predator with its own unique array of costs and benefits. True, there *is* a measure of mechanical equivalence in using a stick to dig up an underground plant, to probe open a termite mound, and to stab a bushbaby hiding in a tree hollow. But, the mechanical equivalence of digging, probing, and stabbing is only superficial, so lumping together each of these disparate food-getting tactics under the same umbrella of "extractive foraging" obfuscates the radically divergent risks and payoffs associated with the different resources being pursued in each case.

The kind of emotional restraint maintained by Fongoli chimpanzees during a bushbaby kill seems an unsustained emulation of the comparatively amplified poise of human hunters, who use much more effective weaponry to maintain even greater distances between them and their much more dangerous prey. In other words, although the

Fongoli chimpanzees are unusually instructive in highlighting the benefits of emotional control for a specific task of great evolutionary consequence—the killing phase of predation—human emotional control is, in contrast, an *essential component* of our makeup. Indeed, human emotional control might be rooted deeply in the "cooperative-communicative context" (that is, using interpersonal communication to initiate and/or sustain cooperation) of our past. It is not surprising that many evolutionary researchers consider the remarkable behavioral flexibility of humans as the ultimate (if somewhat intangible) characteristic that sets us apart from other animals, including even our closest ape relatives: behavioral lability as the essence of humanness. Building on decades of laboratory- and wild-based research on primates and other social vertebrates, comparative psychologists Brian Hare and Michael Tomasello conclude that while chimpanzees and other studied primates lack the ability "to use social cues in a cooperative-communicative context," this capacity is, in contrast, a hallmark of our species and is probably what explains "the unusual behavioral flexibility observed in humans." The cooperative-communicative ability of humans is dependent on two conditions: first, we each need to understand that others of our kind think and have their own needs and motives; second, we need to be able to deliberate about these cogitations of others.

This possession of a "theory of mind," is not, however, a unique capacity of humans. Numerous experiments reveal that chimpanzees also have a theory of mind, but the difference between them and us is that chimpanzees seem to be able to apply the theory consistently only when they are in competitive situations; for instance, chimpanzees are especially clever in deceiving and rooting out deceit in others when hidden food rewards are at stake. What sets people apart from other primates is that we are able to use our theory of mind to interpret and exploit *cooperative* social cues toward the attainment of a common goal that we share with others. True, a chimpanzee can be prompted to collaborate with another chimpanzee in order to solve a problem that, if cracked, holds the promise of a food reward. Though, as Hare explains, the conditions under which this is possible are specific and limited, occurring only

> if (a) the food is sharable, (b) the [chimpanzee] partners are out of each
> other's reach while they [pull the ends of a rope connected to a food tray
> toward their separate, but adjacent, enclosures; the rope slips through a
> u-bolt, leaving the tray out of reach, if only pulled by one chimpanzee at one
> end], and (c) the partners have shared food previously in a similar context.
> If such social criteria are not met, then chimpanzees will not solve the

cooperative problem presented. It seems that subordinate chimpanzees are simply not willing to risk being attacked by intolerant dominants, and dominants are not able to control their aggression toward subordinates trying to obtain food—even if it means the dominants will not receive food either.

Hare and Tomasello thus conclude that the social skills of chimpanzees and other studied nonhuman apes are hampered by their social emotions, those feelings that are produced in the presence of another living being.

As a taxon, humans have, in contrast, gained controlled of our "emotional reactivity" some time during the course of our evolution. Hare and Tomasello have also identified control of emotional reactivity in domestic dogs. Eliminating scent clues that would confound results, the researchers showed time and again that dogs will choose an inverted cup with food under it instead of an empty inverted cup simply because a human experimenter points only at the cup with the concealed prize (figure 4). Chimpanzees fared more poorly than did dogs in this kind of experiment. Likewise, wolves and other wild canids cannot match the abilities of domestic dogs under these experimental conditions. Presumably, throughout the process of domestication, human selection acted to modify the adrenal cortex and limbic system of dogs so that the domestic form now displays reduced levels of fear and aggression as compared to wild canids. Experimental domestication of silver foxes, resulting in the same physiological transformations seen in domestic dogs, seems to confirm this hypothesis.

**FIGURE 4.** Researchers at Duke University's Canine Cognition Center are demonstrating symmetry in the old adage of dog as man's best friend. It appears that dogs feel the same way about us. In addition to the work discussed in the text, the Duke experiments also show that the default reaction of a dog to any human, familiar or unfamiliar, is to trust the person. (Photograph courtesy of Jingzhi Tan)

The historical context under which human emotional control evolved is less clear than that reconstructed for dogs, which occurred much more recently, because dogs have been domesticated within only the past 15,000 to 100,000 years. The sophisticated language abilities of humans might have played a vital role in the process of our gaining control over our passions, but how and when remain elusive. The possibility also exists that emotional control preceded the split between the human and chimpanzee lineages. Recall that the two closest living relatives of chimpanzees—bonobos and humans—are both more docile, cooperative, and just generally agreeable than are chimpanzees. The implication that can be drawn—that humanlike emotional control was a feature of the most recent common ancestor of bonobos, humans, and chimpanzees—is intriguing, but it remains to be tested in any serious way. If true, it would mean that the more extreme emotional reactivity displayed by chimpanzees is a relatively recently *and independently* evolved characteristic, unique among these three closely related primate species. Efforts to increase our presently poor knowledge of bonobo psychology should someday help test this hypothesis.

And, although the data on predation by Fongoli chimpanzees do nothing to shed light on this larger issue, I reemphasize that those data *do* highlight the practical and evolutionary *utility* of emotional control in a primitive hunting context. Hunting weaponry is probably a key factor in allowing that emotional control. Hunting weapons would have provided our prehistoric ancestors with the same advantage—whether or not an overarching, general management of emotional reactivity had yet evolved in hominins by the time they started using weaponry. As mentioned, comparative psychological research on humans, chimpanzees, and bonobos remains to be undertaken that could elucidate the origin of this general emotional management in the hominins.

In the meantime, however, there is *anatomical* evidence from the paleontological record, which suggests to some researchers that the origin of emotional management was well before the invention of hominin hunting weapons. Chapter 2 promised a return to the unexpected sociobehavioral insights provided by analyses of the teeth of the putative Miocene root hominin *Ardipithecus*. Recall that as far back as 5.7 million years ago, *Ardipithecus* possessed relatively reduced, nonprojecting canine teeth and lacked a functioning upper canine honing system. In the context of "typical" ape (living and extinct) morphology, this kind of dentition is unusual. And as such, it seems to demand a fairly grand explanation.

## SEXINESS AND COMMITMENT IN THE MIOCENE?

Anatomist C. Owen Lovejoy rises to the occasion. His model of early hominin sociality links the strangely un-apelike adaptations of bipedalism and canine reduction seen in *Ardipithecus* to the inter- and intrasexual relationships of this purported earliest human ancestor. Lovejoy first notes that "giving up" the large canines of the type possessed by its ape relatives was a sacrifice for male *Ardipithecus*. First, the lack of big, dangerous fangs severely compromised the offensive and defensive capabilities of male *Ardipithecus*. Second, the short, stubby canines of *Ardipithecus* were also useless in aggressive displays against other male competitors. By extension, mate selection could not and did not involve such displays and probably lacked significant antagonism between males. For such a radical departure from the typical ape condition to take hold, it seems all but certain that *Ardipithecus* females were preferentially selecting nonaggressive males as mates. Modifying the anthropological oldie of "meat-for-sex," Lovejoy ventures further. Lacking any avenue of brutal directness to a desired female, the reproductively successful male regularly provisioned a female partner with food (not necessarily meat, but at least some kind of high-protein and/or high-fat food), favors she returned in kind with sex. Additionally,

> [males] would benefit from enhanced male-to-male cooperation by virtue of their philopatry [Lovejoy assumes that, as is the case for most other African apes, *Ardipithecus* and other hominin males stayed together in the groups into which they were born], because it would improve not only their own provisioning capacity, but also that of their kin. Foraging could be achieved most productively by cooperative male patrols. . . . Provisioning would reduce female-to-female competition by lowering reliance on individual "sub-territories" (as in chimpanzees) and/or resource warding and would improve (or maintain) social cohesion.

In this model, the emergence of hominin bipedalism is explained by the need of males to carry food with which to provision females. A female, in turn, ensured loyal, uninterrupted provisioning from her pair-bonded male by becoming continuously sexually receptive to him. Most female nonhuman primates are sexually receptive only during estrus, at or around their times of ovulation, when that receptivity is advertised to males by quite apparent swelling of their perinea (think of the angry red butts of zoo monkeys that elicit so much guffawing by second graders—and their parents). In contrast, human females—lacking visible clues to their times of ovulation—keep us males guessing, and we are thus unsure for any given sexual encounter with a woman whether it will result in successful reproduction. Because of this, Lovejoy posits that female hominins developed reproductive crypsis in order to ensure unremitting sexual interest from males seeking to increase their Darwinian fitness through multiple successful conceptions and births of their offspring. Lovejoy contends further that the development of female reproductive crypsis is firmly rooted in our past, at least by the time of *Ardipithecus,* nearly 6 million years ago. If he is right about that, and about the accompanying package of humanlike sociobehavioral organization of *Ardipithecus,* then, by extension, conjecture of a deep prehistory for general emotional control in hominins is also reasonable.

However, it is important to remember that there is still no consensus that *Ardipithecus* is, in fact, a genuine hominin. Paleoanthropologists Bernard Wood and Terry Harrison are the latest skeptics to stress that humanlike features (such as non-honed canine teeth, a forward-placed foramen magnum, and hip bones whose upper portions are short and broad)—argued to be unique to be *Ardipithecus* and later hominins—are also found in other Miocene ape species, some of which are even more ancient than is *Ardipithecus.* To Wood and Harrison this means that the hypothesis of *Ardipithecus* as the direct ancestor of all later hominins is still unresolved. Other Miocene ape specialists, like David Begun, accept that *Ardipithecus* was a hominin, but they also contend that it might more specifically represent "a surviving relict of an early branch of the hominins without a direct relationship to later taxa [like *Australopithecus* and *Homo*]."

Thus, although holding powerful potential to speak about the hominin condition millions of years ago, the paleontological record is still unclear in ways that currently hinder realization of that potential. This is not to say that these interpretive gaps are unspannable—but certainly much more work is needed to clarify the evolutionary relationships of the sundry species that have so far been identified. In the meantime, as might be predicted, the complementary archaeological record is poised to contribute to the more specific question of hunting by hominins who were in control of their emotions—regardless of when that control was achieved. If animal bones recovered from archaeological sites preserved impact damage from spears or arrows, then this would obviously indicate that hominins possessed distance-providing hunting technology. However, the earliest cultural record is not so forthcoming. And thus, the motivated researcher must rely on much more subtle clues when probing the beginnings of humanlike hunting. It is altogether fitting that questing that kind of faint paleoanthropological evidence is traced back to Raymond Dart and to his provocation of a most understated and meticulous young man.

# 4

# Conceiving Our Past

Thy rod and thy staff they comfort me.

—Psalm 23:4, *The Bible (Authorized King James Version)*

In 1955, when he first encountered Raymond Dart, Charles Kimberlin Brain had already been a different person for twenty-two years; at the age of two, Brain announced to his parents that he should be called Bob from that moment forth—the name Charles resonating "much too grandly" for his taste. And, Bob Brain is still Bob Brain today, eight decades later. Outwardly unassuming, Brain's childhood pronouncement betrayed that he is built around a solid core of self-assurance—the stuff that would on that day in 1955, at a conference on prehistory in Livingstone, Zambia, move him naturally to question Raymond Dart's just-completed defense of the "killer ape hypothesis."

Brain was just starting his dissertation research on the geology of the South African ape-man caves. But, with a recent paper on stone artifacts from the Makapansgat site published in *Nature* (then, and now, the world's most prestigious scientific journal), his star was already rising rapidly. With that kind of juice, Brain would surely be able to soon mount a proper investigation of Dart's jarring pronouncements about an *Australopithecus* death-culture. But, Brain's life soon twisted and turned. Following his discovery of the Makapansgat stone implements, he made an even more important discovery of the first known primitive quartzite tools from Sterkfontein Cave. If Brain imagined that his advancement should follow rapidly, John Robinson (chapter 2), the site's well-connected director, the then-Curator of Palaeontology at the Transvaal Museum, *and* also Brain's doctoral dissertation supervisor, thought otherwise. Robinson proved to be a glory hound who was obviously threatened by Brain's moxie. He also proved to be a bully who forbade Brain to publish anything on his Sterkfontein discovery, commanding that he, Robinson, would instead be the one to publish the findings and their implications. If Brain disobeyed, Robinson ensured him that he would use all is considerable influence to deny Brain a career in paleontology. Undaunted, Brain defied his tormentor's edict (Bob tells me that "Stuff you!" was his exact reply to Robinson), and announced his discovery of the Sterkfontein artifacts in his 1958 doctoral dissertation. Brain's play, while just and dignified, brought Robinson's promised wrath: he was barred from working on all hominin sites in Robinson's control (in other words, nearly all of the sites then known in South Africa). More, at the Transvaal Museum Brain was banished to the position of Curator of Lower Vertebrates, where he was left to research amphibians and reptiles for several years. Then, in 1961, Brain accepted an invitation to become Deputy Director of the National Museums of Rhodesia and was integral in setting up several new natural history museums in that, his native, country. It was not until 1965, when Robinson finally took leave of his position at the Transvaal Museum, that Brain returned to replace his antagonist. At long last, he would now be able to launch the first truly empirical test of Dart's fantasy that humanity was sprung from the bosom of a primordial killer ape.

Brain's approach to reconstructing early hominin behavior was as exacting as Dart's was unrestrained. Brain was also able to go beyond Dart's reliance on the site of Makapansgat in order to evaluate the capabilities of *Australopithecus*. His new position as Curator of Palaeontology at the Transvaal Museum now gave him access to the large fossil assemblages from Sterkfontein and Kromdraai caves, and he even began his own excavations at the site of Swartkrans. Moreover, because of his long experience as a naturalist, Brain approached the analysis of ancient material remains in a way that was radically distinct from that of Dart and his contemporaries. Dart was cyclonic, relying almost solely on his frenetic imagination to reconstruct ape-men as primeval felons, whereas Brain took to heart the patient axiom of two pioneering geologists—that the present is the key to the past.

Like James Hutton and Charles Lyell long before him, Brain fully appreciated that to observe, *and thus decipher,* cause-and-effect relationships in the modern world was the tool that could render prehistory explicable. It is *behavior* that the paleoanthropologist is after, it is lifestyle, ecological interaction: dynamics. But, the archaeological and paleontological records—the databanks for reconstructing the lives of extinct hominins—are phenomena of the now, composed only of the *static material consequences* of past behaviors and events, which are, of course, directly unobservable today. James Hutton understood this epistemological quandary way back in the late eighteenth century. More important, he also conceived *actualism* as a solution to overcome it—a method to span that seemingly unbridgeable chasm between the goal of reconstructing ancient actions and our inert, contemporary sources of data. But, alas, Hutton's thesis on actualism is buried deep within the two thousand–plus pages of his 1794 three-volume opus, *An Investigation of the Principles of Knowledge and of the Progress of Reason, from Sense to Science and Philosophy.* Even the Enlightenment's heightened tolerance for pedantry had its limit and Hutton's incoherent, tortuous prose easily exceeded it. And so it was that one of the historical sciences' greatest ever heuristic advances went generally unappreciated upon its initial publication and for nearly half a century beyond.

It was only in the early 1830s that Charles Lyell, Hutton's fellow Scot and great advocate, was able to break through and convince the intelligentsia of actualism's unmatchable utility. Lyell's classic treatise, *Principles of Geology,* lays out clearly the relationship between actualism's foundational assumption, that natural laws have not and do not change across space and time, and its implication, that this invariability of natural laws means that phenomena formed in the past did so through processes still operating today. Those past processes—ancient dynamics—are thus knowable to us through the study of contemporary processes. The prehistorian is thus provided a workable method to disclose ancient processes: observe natural events, conduct controlled experiments, and establish cause-and-effect relationships between a dynamic process and its material consequences—all today, in the modern world. If the material consequences of these observed processes match material remains from fossil and archaeological sites, the prehistorian then positions himself to use analogical reasoning to say something about human behavior in the past.

Brain was one of those wholly convinced by the pragmatism and explanatory power of actualism. Indeed, he was really the first paleoanthropologist to fully appreciate that understanding bone-accumulating processes in the modern world was the only hope to unriddle the formation of what Dart so vehemently argued were the charnels of prehistoric ape-men. On a more personal level, Brain found the prospects of actualism and the whole enterprise of its undertaking immensely appealing.

Studying the transformation of living things first into piles of bones and then, ultimately, into fossils was not just pioneering. It was also embracing, edgy work, well suited to the rugged white African individualist, like Brain, who was not yet so uncommon in the mid-twentieth century. Sometimes joined by his friend, a jocular Namibian farmer named Attila Port, Brain poked about bloated carcasses and moldering bones. He and Port confronted jealously feeding carnivores for a first-hand look at their meals, and they observed the livestock-slaughtering practices of tribal people—all the while suffering the common deprivations of African fieldwork: sickness, heat, and (often self-inflicted) personal peril. The son of German immigrants to Namibia (then, Deutsch-Südwestafrika), Port grew up quickly and grew up tough. His father once armed five-year-old Attila with a thorny tree branch and sent him into a leopard den to confirm their suspicion that it was occupied by a mother and her cubs. "Equipped with a torch and a .450 caliber revolver, the father followed. As was to be expected, the leopard charged the intruders, and Mr. Port shot it over his son's head. After they dragged the leopard from the lair, the boy crawled back in and brought out the two cubs." Given that upbringing, it is hardly surprising that forty years later, when Brain wanted to explore a leopard den on Attila's property, the farmer handed him a zebra leg bone, offering the laconic advice to shove it in the leopard's mouth should one spring out of the darkness. Bucking up, Brain took the bone and entered the den, edging his way past a still-steaming pile of disemboweled guts. But, upon hearing a throaty emission from the deeper recesses of the den, Brain prudently reversed out at a rapid clip, deciding further investigation could wait until another day when the den was unoccupied.

The misadventure on Port's farm was just one manifestation of Brain's long and storied association with leopards. More relevant to this narrative, it was Brain who first noticed that the fossilized skull of an australopithecine child from Swartkrans was injured by two through-and-through punctures—the spacing of which match exactly the span of the canine teeth of a leopard's lower jawbone, which was also recovered from the same deposit (figure 5). That observation became the loudest ringing nail driven into the coffin of the "killer ape hypothesis." But, it was only one of those nails, as Brain's comprehensive and deep interment of Dart's theory proceeded unflinchingly throughout the 1960s, '70s and '80s. Brain's research on bone-accumulating processes in modern caves revealed that the patterns of animal body part representation and bone damage documented by Dart at Makapansgat, and by himself at

**FIGURE 5.** The skull fragment of an ape-man child from Swartkrans Cave, South Africa, with two punctures (arrows) created by the lower canine teeth of a leopard whose jawbone was also found at the site. (Photograph courtesy of C. K. Brain)

Swartkrans, Sterkfontein, and Kromdraai, were comfortably attributable to forces *other* than the supposed predatory and murderous behavior of cave-dwelling ape-men. Portions of antelope bones represented at these sites were not there because they were selected by australopithecines to use as tools and weapons. Rather, they were simply those skeletal parts that were too dense or too poor in nutrients for carnivores to destroy when consuming the carcasses that they had dragged into the caves. Like modern caves, the ancient sites could no longer be reconstructed as the dank domiciles of ape-man. Instead Brain revealed them as static receptacles of feeding debris that was generated over long periods by large cats and hyenas (see chapter 2). And, based on the great number of their fossils, including many scarred by carnivore tooth marks, australopithecines were apparently one of the favorite prey animals of those predators. Dents and missing sections of primate skulls were not created by bone clubs, but instead by steadily accruing pressure as bones were being buried and compressed by sediment that washed into the cave year after year. Other traumatic damage to bones was caused by high-impact geological events, such as great chunks of rock spalling off cave roofs and crushing bones that rested on their floors. The reason that skulls of ape-men and monkeys predominate the fossil samples is because heads offered little in the way of nutritional value to carnivores; the rest

of a primate body, a comparative wealth of meat and marrow, was consumed in its near-entirety. In the process, bones were demolished into
unidentifiable splinters. As Brain continued to develop these thoroughly
convincing, but comparatively mundane, rejoinders to Dart's fantastic
scenarios of how the ape-man sites formed, the "killer ape hypothesis"
was methodically ground away, and the field of taphonomy—the study
of processes affecting bones after an animal's death—took tight hold.

## A GENTLER NOBLE SAVAGERY?

Far from being the cannibalistic alpha predators of Pleistocene Africa,
Brain now contended ape-men were instead the insignificant quarry
of large cats, wretched creatures, quaking in the gloomy margins of a
carnivore's world. Brain recognized immediately the implications of
his work—published collectively in 1981 under the provocative title
*The Hunters or the Hunted?*—and agonized about having to relay it
all to Dart (figure 6). The older man had always been encouraging

**FIGURE 6.** Bob Brain at Swartkrans Cave, South Africa, where his long-term research
changed the prevailing views, propagated by Raymond Dart, of ape-men as mighty
hunters and misanthropic killers. (Photograph courtesy of Jason L. Heaton)

to Brain, and the two developed a close fondness as Brain's career accelerated and Dart's began a slow decline. In Brain's own post hoc evaluation, though, he had less to worry about than he originally fretted: "As the implications for his cherished ideas about early [hominin] hunting prowess and aggression became clear to [Dart], he fell silent for a few minutes. Then he became more and more enthusiastic, saying, 'This is wonderful! At last we are getting closer to the truth!' Within a week he had nominated me for an award. Such generosity of spirit was wonderful to see and showed that Dart was far more interested in the subject itself, than in the promotion of his own ideas."

Promotion of ideas was, however, soon to follow Brain's shake-up of the way in which we perceived our ancestors. As Brain was conducting his taphonomic studies, other scientists were carrying out nascent work on social organization in the few groups of hunting and gathering people left in the world in the late 1950s and early 1960s. The most high-profile of this research was on the Kalahari San (formerly known as Bushmen, a term many people now consider pejorative), the same diminutive, bow-and-arrow-toting people who starred in the oddball *The Gods Must Be Crazy* films. Recent genetic research shows that some San groups are the oldest known populations of modern people in the world today. But, their traditional lifestyles were already in rapid transition across southern Africa—set in motion by the outside pressures of neighboring African farmers and herders, and by the economic and militaristic forces of the political states in which they resided—by the time Richard Lee, and other important modern-era ethnographers of the San, began documenting their culture in the mid-twentieth century. Still, observing the remnant traditional societies that *did* exist among the San, anthropologists recognized that they were organized socioeconomically as are many other tropical and subtropical foraging groups, with women concentrating on plant gathering and men doing the bulk of the group's hunting—a sexual division of labor.

A theoretical growth of the ethnographic work on the San and other hunter-gatherer groups coalesced under the banner of "man the hunter," the influential anthropological paradigm that, like Dart's "killer ape hypothesis," promoted hunting as a prime mover in human evolution. In large part because of ethnographer Elizabeth Marshall Thomas's 1959 characterization of them as the "harmless people," a stereotype proliferated as accepted wisdom that the San and other hunter-gatherers were essentially peaceable folk, contrasting starkly with hyperantagonistic industrialized Westerners. Eventually, a wealth of field data belied this perception. Take, for instance, the observations of Richard Lee, who, over the course of five years in the early 1960s, documented more than thirty serious fights among the San whom he observed. Lee also brought to light the records of twenty San-on-San murders in the groups that he studied. So, once again—this time under the banner of "man the hunter"—human predation was paired causally with aggression, a point that did not escape the insightful commentator who noted an "attendant misanthropy" in all hunting hypotheses of human origins. "Man the hunter" also put great emphasis on the fact that hunting is typically a male endeavor and that males are generally more aggressive than are females. With "man the hunter's" reinforced emphases on maleness, hunting, and aggression, the role of women in shaping our evolutionary trajectory toward full humanness was once again relegated.

But, in direct opposition to the implications of "man the hunter," and simultaneous with that paradigm's formation, as well as with Brain's field studies, a third, much more broadly affecting phenomenon was materializing during the late 1960s and early 1970s. And, so science was now on a collision course with the escalating social movement of feminism. It is a supreme irony that Brain's conclusions about the meekness of human ancestors, based on conscientious field research and carefully drawn inferences, was the crack in the floodgate through which a wave of pseudo-science began to pour. Feminist researchers were soon as guilty as were adherents of "man the hunter" in carelessly applying analogy from the modern world to explain the past. Because it supplies a more dependable daily source of calories than does hunting, plant collecting, primarily women's work in many foraging societies, was now touted as the key human adaptation. For advocates of "woman the gatherer," "man the hunter" had it all wrong: prehistoric women were the crucial sex in our becoming human. With that perspective, and knowing that human females are also typically less aggressive than are males, the evolutionary importance of violent propensities in our ancestors was downplayed. Brain's research was unfairly pulled into the argumentation as support for a new "make love, not war" model of early human lifeways. Australopithecines and early *Homo* were now perceived as much like the neutered, pre-Goodall stereotype of the living apes—gentle vegetarians, doing all they could to evade the menacing daily overtures of large carnivores, and with no aspirations to become predators themselves.

Detect the activism that elbows its way into a summation of the "woman the gatherer" reactionary movement: "['Man the hunter'] promotes the idea of male aggression as necessary for hunting and for protecting the weak and passive females and children and assumes male dominance over females inherent to the hunting way of life." There is, no doubt, considerable perception in this statement. It correctly recognizes the link forged between aggression and hunting by (like Dart and Ardrey before them) the proponents of "man the hunter." But, the quote also reveals itself as more the product of some kind of extended postmodernist critique than the outcome of data generation, deductive reasoning and careful hypothesizing. And so it went after Bob Brain's amoral taphonomic results: indignant, politically colored pronouncement followed indignant, politically colored pronouncement—all in the near-absence of direct study of the fossil and archaeological records.

**BACK TO THE EARTH**

The flame of paleoanthropological empiricism was not, however, completely extinguished. By the mid-1970s, the work of archaeologist Glynn Isaac was providing exciting new field data from the Koobi Fora region, on the edge of Lake Turkana in northern Kenya. Isaac's excavations of numerous archaeological sites throughout Koobi Fora revealed evidence of hominin behavior that, at the time, was some of the oldest known. The Koobi Fora localities—most of them estimated to be nearly 2 million years old—showed a consistent intersite archaeological pattern, with abundant stone artifacts found in direct spatial association with the broken fossil bones of large vertebrate herbivores. Allowing for the vast technological superiority of modern hunter-gatherers, those ancient archaeological associations are roughly analogous to the comingling of discarded material culture remains and prey bone residues observed in the abandoned campsites of San and Hadza foragers.

The modern occurrences—collective results of cause-and-effect, action-and-material consequence relationships—are instilled with myriad behavioral significance and inferable social dynamics. At a fundamental level, ethnographers and archaeologists are right to see them as marking humanly meaningful, central places on the landscape—places to which men and women returned daily to share the different foods that they had collected and to act out basic rites of domesticity and sociability. In order to explain the archaeological patterning of materials at the Koobi Fora sites, Isaac inferred similar behaviors for our Pleistocene ancestors. Under his "home base model," Isaac argued that by at least 2 million years ago, pair-bonded male and female hominins, each day, left a fixed spot on the landscape—a camp or home base—foraged independently or in same-sex groups (presumably for meat and plants, respectively) and then returned to the same home base to share their collected food with their mates and to interact in a very humanlike way. Thus, even though Isaac's approach was substantively novel in "inferring up" from an empirical base, his "home base model" was, in another significant way, like "man the hunter" and "woman the gatherer" before it. Its construction relied on analogy with modern foragers to explain the organization of protohuman society.

Isaac maintained his dualistic approach—feet planted firmly on the concreteness of excavations and their contained material residues, while, at the same time, remaining the theoretician—as he recruited a small band of graduate students to test and hone his "home base model." Among them, Henry Bunn was charged with analysis of butchery damage—defleshing cut marks, where the sharp edges of stone tools accidentally sliced through meat into underlying bone, and percussion marks, created when hominins used river cobbles to pummel open and harvest the marrow of animal leg bones—on the animal fossils from the Koobi Fora sites. Combining the results of this analysis with his additional identification of abundant butchered bones from contemporaneous sites at Olduvai Gorge, in Tanzania (figure 7), Bunn concluded that, by 1.8–1.5 million years ago, hominins were quite capable meat foragers, gaining regular access to ungulate carcasses that were large and complete enough to share in the way envisaged in Isaac's "home base model." For instance, at a single site in Olduvai Gorge, FLK Zinj, Bunn documented the remains of nearly fifty large mammals and cut marks on more than two hundred fossil pieces broken from bones that once supported copious amounts of flesh. Many of the FLK Zinj fossils also bear hammerstone percussion marks.

Bunn's conclusions that this collective evidence indicated hominin meat eating, and, by extension, probably hunting and/or aggressive scavenging of large ungulates, were soon challenged by an imposing figure in the pantheon of contemporary American archaeology. Lewis Binford was an architect of the New Archaeology, which, beginning in the 1960s, shifted the discipline toward a more rigorous application of scientific methods and use of deductive reasoning and hypothesis testing. Sporting an establishment haircut, and a loudmouth opinion on seemingly every aspect of the human past, Binford wrote profusely (and often viciously) on subjects as varied as the foraging adaptations

FIGURE 7. Close-up of an antelope leg bone from the site of FLK Zinj, at Olduvai Gorge in Tanzania, which shows multiple cut marks created accidently by a Pleistocene hominin who used a sharp-edged stone tool to butcher its meat (scale in millimeters). (Photograph courtesy of Henry T. Bunn)

of modern Arctic natives and the prehistoric origins of agriculture in the Near East. His 1981 book, *Bones*, was a tirade against what he saw as the modern mythologizing of early hominins by "professed" scientists who should have known better. A specific objective of *Bones* was to dismantle the notion that hominins were ever efficient hunters until only very recently. Even Neandertals—a late-occurring, seemingly advanced species, who possessed brains larger than those of modern humans and who survived exceptionally well in the grinding extremes of Ice Age Europe—were relegated by Binford to the status of bumbling, moronic scavengers.

So, it is not surprising that Bunn's assertions about the much older fossils from archaeological sites at Koobi Fora and Olduvai Gorge did not escape the elder statesman's ire. In a postscript to *Bones*, Binford categorically dismissed Bunn's analysis, sniffing that it was "a feeble attempt to salvage or save the conventional interpretation that all the . . . [fossils] were deposited . . . by the [hominins]." Indignation aside, Binford was on empirically shaky ground: Bunn had studied the fossils in question, Binford had not.

Bunn responded with a salvo to open his ugly public contest with Binford. Their debate over early hominin hunting would play out in the pages of anthropology journals and books throughout the 1980s. In retrospect, at a distance from its entertainment value, Binford's huffing is laid bare as just that—hyperbole, pure and simple. Bunn's data were solid. For all his deification by modern archaeology, Binford's most important contributions will remain largely theoretical. His role in building a coherent picture of what happened in prehistory has had much less impact. The only justifiable challenge to Bunn's conclusions would have to come from someone else, someone who was at least willing to go to Africa and look at the fossils in question.

## KING OF THE BEASTS: A MUNIFICENT RULER?

A new empiricism in studies of hominin foraging, spurred by the taphonomic revolution headed by Brain, Isaac, Bunn, and Binford, assured that that challenge was just around the bend. It was led by a young researcher named Rob Blumenschine. By the late 1980s, Blumenschine's observational and experimental work in Tanzania's Serengeti ecosystem had convinced a majority of his colleagues, as well as the informed lay community, that the earliest hominin meat eaters were almost exclusively scavengers. This conclusion was in many ways no different than that of Binford's—but greater credibility was given to Blumenschine's assertion, largely because he applied sophisticated ecological theory in his field studies. This, combined with his scrutiny of fossil bone surfaces for marks of carnivore and hominin feeding (and, in addition to corroborating studies by his students and collaborators), led to touting of a "passive scavenging" phase in early hominin evolution.

Central to Blumenschine's version of the "passive scavenging hypothesis" is the understanding that most early archaeological assemblages are palimpsests—the combined result of the activities of not just hominins, but also of large carnivores, bone-collecting rodents and birds, and burrowing worms and vermin (as well as a multitude of geological and chemical processes). Appreciation of the heterogeneous nature of the archaeological record is not new. Even Raymond Dart—who concocted, nearly of whole cloth, an ape-man osteodontokeratic culture to explain dramatically otherwise prosaically explainable patterns of bones and bone damage at Makapansgat—recognized that the bone-collecting proclivities of other animals could prove fatal to his hominin-centric story of site formation and to the "killer ape hypothesis." The nagging menace of this possibility so consumed Dart over the years that he reported an extraordinary claim in his quizzical 1956 paper, entitled "The Myth of the Bone-Accumulating Hyaena":

> [We] have . . . made repeated efforts to locate some of these presumptively bone-littered dens of hyaenas and other carnivores. . . . Despite the prevalence of hyaenas and other large felines throughout Africa these supposed carnivore bone accumulations have proved most elusive. . . . Even a vigorous press campaign conducted by Mr. Alun R. Hughes failed to elicit definitive evidence leading to the localization of any such postulated cavern deposit. . . . In 1953, therefore, Mr. Hughes approached the Kruger National Park administration for permission to excavate brown hyaena lairs. . . . In neither of the two hyaena lairs then investigated, which were both used by hyaenas at the time of his visit . . . did they find any bones whatsoever; nor was there any accumulation of bones in the vicinity of the two lairs.

This remarkable protestation from Dart flies in the face of a mountain of research by wildlife biologists and paleoanthropologists, who have routinely observed hyenas collecting carcasses and bones and who have excavated bone-rich hyena dens from all over Africa and the Middle East.

Blumenschine's particular take on the taphonomic potential of carnivores is more in line with the ecological reality that carnivores are, in fact, common taphonomic agents. But, in its own way, it is no less stanch than were Dart's bizarre denials. Blumenschine contends that those African archaeological sites relevant to deciphering early hominin meat foraging—like especially FLK Zinj, at Olduvai Gorge—are *solely* the result of the serial but *interdependent* activities of multiple carcass consumers. In particular, Blumenschine sees a recurrent taphonomic pattern in several important Stone Age bone collections, indicating to him that the assemblages formed, nearly invariably, through the actions of hominins scavenging carcass residues that had been abandoned by lions, the primary predators of ungulates.

This interpretation flatly contradicts the reasoned and rigorously supported findings of Henry Bunn, who instead concluded—based largely on the abundance and distribution of cut marks on meat-bearing bones—that the large ungulate carcasses represented in sites at Koobi Fora and Olduvai Gorge were accumulated primarily through the actions of hominins with exclusive or, at the least, very early access to those carcasses. Unlike Bunn, Blumenschine mostly downplays the FLK Zinj cut mark evidence and is instead particularly impressed by hammerstone percussion damage on some of the site's fossil bones. Blumenschine elevates the evidence of percussion damage in the belief that it indicates a defleshed condition for carcasses when hominins were able to safely scavenge them. Blumenschine also places great stock in what he *considers* an abundance of carnivore tooth marks on many of the FLK Zinj bones (more later in the chapter on problems with this perception).

Because lions play the central role in his model of archaeological site formation, Blumenschine (and other researchers) followed in the pioneering footsteps of Bob Brain and recorded exactly how the modern big cats of Africa dismantle carcasses when they first gain access to them. Lion feeding is, of course, variable across time and space, and various research groups have observed different degrees of carcass destruction in the courses of their studies. Sometimes lions are especially hungry; other times not so much. Some prey carcasses are consumed by large prides, under intense competition; others under little contest, by smaller groups or single individuals. And, a host of other variables—even those as seemingly idiosyncratic as the time of day, ambient temperature, availability of shade, and the density of swarming flies—have the potential to affect significantly how much of a carcass lions (and other predators) will consume. However, based on a robust observational dataset, a general pattern emerges in which, all other factors being equal, lions usually deflesh carcasses nearly entirely and only rarely break bones to extract marrow. Because lions are big, they are able (especially in concert with other members of the group) to dispatch large prey, usually ensuring that the carcass's meat, alone, will sate the whole pride. In addition, as is true of all cats, the highly specialized, meat-slicing teeth of lions are not well adapted to chewing bones. Unlike those of a dog or hyena, cat teeth are relatively delicate, susceptible to breaking if brought into forceful contact with a hard item like bone. This fact well explains the generally complete, unbroken condition of the bones of carcasses eaten by lions.

In the light of these observations, it is apparent that hominins who scavenged passively from lion kills would have typically only had access to the neglected within-bone nutrients of carcasses: brains encased in skulls and marrow encased in long leg bones. In fact, many pertinent archaeological assemblages from Africa preserve an abundance of hammerstone percussion marks. This taphonomic evidence means that hominins did, indeed, harvest marrow from long bones.

The final touch in Blumenschine's model of archaeological site formation accounts for the rarity at sites like FLK Zinj of the articular joint ends of ungulate leg bones. Blumenschine suggests that hyenas engaged in a third stage of scavenging, consuming the ends of prey leg bones, which were still laden with edible grease when they came upon them. Blumenschine elaborates that the greasy bone ends were ignored by hominins during their post-lion scavenging activities because they still lacked, at that early stage of prehistory, a technology for boiling. Boiling the bone ends could have rendered the grease digestible for hominins. In contrast to hominins, hyenas don't need technology to free and metabolized bone grease; they have their robust teeth, powerful jaws, and specialized digestive physiology to get those jobs done. In sum, Blumenschine's three-stage model sees early hominin meat-eating as bracketed by the activities of other mammalian carnivores: first lions killed and ate what they wanted of prey animals; hyenas batted clean-up, eating the greasy ends of leg bones; and, hominins were stuck in the middle, avoiding conflict, passively scavenging scraps of brains and marrow.

Never mind that a subsistence strategy based exclusively on passive scavenging is completely unknown among other mammalian meat eaters (even hyenas—the poster species for scavenging exclusivity—are now known to be avid and successful hunters, their maligning by ridiculous humans thus unfair on two levels: first, despised as scavengers as if scavenging was some kind of moral deficiency, and, second, so prejudged long ago without adequate behavioral data that, when eventually accumulated, obliterated the perception). Everything finally seemed to add up, and textbook writers, as well as researchers, contentedly massaged disparate lines of evidence into a cohesive new story of our evolution. It never made much sense to a lot of people that early Pleistocene hominins—who in adulthood only reached the size of a modern nine-year-old child, who lacked claws and sharp teeth, who were slow footed, and who possessed only the most rudimentary technology—could have operated effectively as predators. A model of passive scavenging for *possibly* ape-men (Robinson's "dietary hypothesis" still casted a long shadow of doubt that *Australopithecus* ever ate any meat at all), and surely for the earliest meat-eating members of the genus *Homo*, was at balance with this perception. The brilliance of Blumenschine's model was its reassuring congruency. It accommodated the presence of cut marks and percussion marks, evidence of hominin butchery that was clearly apparent in many early fossil assemblages, but stopped short of radical claims for humanlike capabilities in primeval hominin species, including *Homo erectus*, Dubois's "man among ape-men." By the late 1980s, the passive scavenging model held top billing. Science dictated it, so asserted data generators like Rob Blumenschine and his students, and so promulgated popularizers and professors alike. Home bases and hunting devolved into so much quaintness, embarrassing anachronisms touted by academic dinosaurs.

## BOB AND WEAVE

There is, however, a problem with this pat dismissal of early hominin hunting capabilities. It comes down to cut marks. Cut marks—their abundance in early archaeological bone assemblages and their high frequency on once-meaty ungulate bones—have always been thorny for proponents of passive scavenging. If they even bother to confront the cut mark evidence, these partisans typically explain it away, arguing that hominins made the marks when removing tiny scraps of flesh and sinew left adhering to bones after lions had had their fill of meat and abandoned carcasses. In this view, hominins were so abject that even desiccated, worked-over morsels were worth the tedious effort to remove them. Or, maybe it was that hominins scraped away scraps of meat in order to make it easier to pound open bones to get at the marrow contained within them.

Paleoanthropologist Manuel Domínguez-Rodrigo tested these challenges proposed by Blumenschine and his colleagues. Domínguez-Rodrigo's observations of twenty-eight modern ungulate carcasses fed on by wild lions in Kenya corroborated those of others, who observed that very little meat is left on a carcass after lions have finished eating. The scraps that did remain on the carcasses studied by Domínguez-Rodrigo were insignificant, with most stuck in the nooks and crannies of the neck vertebrae. A few bits of soft tissue adhered to leg bones, but none occurred on the middle shaft portions of the upper leg bones (the humerus at the front of the body, and the femur at the back of the body), which in life are heavily muscled. That observation is especially relevant because the middle shaft portions of upper leg bones are among those skeletal parts mostly commonly cut-marked in important archaeological assemblages, like that from FLK Zinj. Logically, there would have been no reason for early hominins to have cut-marked these bone portions had they come across the bones only *after* they were completely defleshed by lions. Putting together these ideas "helped to contextualize the relevance of cut mark location. Cut marks observed on those sections of bones that are completely defleshed by primary carnivore consumers [such as lions] likely reflect early access to carcasses by hominins, a behavior that is inconsistent with passive scavenging." Thus, the hammerstone percussion marks also observed in many of these archaeological samples are not the simple reflections of late access, "marginal" scavenging, restricted solely to marrow harvesting. Instead, the percussion marks, in combination with the profusion of cut marks, indicate thorough processing of nearly complete carcasses by hominins, which they first defleshed *and then* demarrowed.

None of this elaboration sunk the "passive scavenging hypothesis," however. Its proponents are dogged and present clever ideas to accommodate the patterns of archaeological cut marks within their models of late access to carcasses by hominins. Among the plausible hypotheses offered is the notion that hominins could have obtained fleshed carcasses of medium-sized antelopes in the wake of catastrophic events, such as the seasonal mass drowning of wildebeest herds, which occur today in the Serengeti and other African ecosystems. Hominins stealing small antelope carcasses, killed then cached and temporarily abandoned in trees by leopards, is another idea; this one seeks to explain why the bones of smaller ungulates from sites like FLK Zinj show abundant butchery marks on preferred, once-meaty parts of carcasses. Others argued that when sabertooth cats abandoned the very large herbivores that they killed and ate, the carcasses would have still retained good amounts of flesh for passively scavenging hominins; the cut marking seen on the archaeological bones of these large animals was thus completely expected and wholly explainable.

Like all scientific hypotheses, these that sought to balance the *reality* of ancient cut marks with the *idea* of passive scavenging generated testable predictions. And, time and again, they failed their archaeological tests. In failing, they also effectively falsified the overarching hypothesis of passively scavenging hominins. For instance, counter to predictions of the "mass-drowning hypothesis," relevant fossil assemblages from sites at Koobi Fora, Olduvai Gorge, and elsewhere are composed of entirely different ranges of species and age distributions of dead animals than are observed in modern mass drowning events. Contradicting forecasts of the "tree-stored-leopard-kill hypothesis," remains of small antelopes from the Olduvai sites show cut marks on bones that once supported large packages of meat. Based on modern observations, we know leopards prefer to eat these same portions immediately after they make a kill—so, by extrapolation, cut marks on those bones indicates that they must have been muscled when encountered by hominin consumers, and *not* first partially consumed by leopards. Further, analysis of a prey bone assemblage from a sabertooth den indicates, in contrast to the "sabertooth hypothesis," that those extinct cats were quite efficient carcass defleshers. The bones of sabertooth prey are covered with tooth marks, and so it is apparent that a hominin scavenging passively from their feeding residues would have been confronted with meatless bones.

However, most damning to the "passive scavenging hypothesis" is a bone-by-bone reanalysis of fossils from FLK Zinj that Blumenschine diagnosed were tooth-marked by carnivores. Through application of their findings on experimentally controlled bone decay, Domínguez-Rodrigo and his colleague, Rebeca Barba, showed that the incidence of carnivore tooth marks at the site is in reality far lower than claimed by Blumenschine. Many of the marks on the FLK Zinj fossils indentified as tooth marks by Blumenschine are actually grooves created by biochemical erosion of bone surfaces. Under the right conditions, when meatless bones are deposited on the ground, colonies of fungi and bacteria form on them. These microbial formations excrete enzymes and acids in order to metabolize the osseous tissues on which they are adhered, wearing away layers of the bone surfaces in patterns that can superficially resemble carnivore tooth marks. But, unlike typical tooth marks, these biochemical marks are usually irregularly shaped, eccentrically trending in linear dimension, and grouped in dendritic patterns—all shapes associated with the growth of fungal and bacterial colonies and not with carnivore tooth marks.

Beyond these taphonomic falsifications of the "passive scavenging hypothesis," another tantalizing line of evidence from Olduvai Gorge seems to support the hypothesis that early *Homo* was primarily a hunter rather than primarily a passive scavenger. It was at the 1.5 million-year-old site of SHK that a team led by Domínguez-Rodrigo and Fernando Diez-Martín recovered the cranial fragments of a hominin child, who was about two years old when he died. The fossil is blemished by porotic hyperostosis, a bony pathological condition that in modern people is associated closely with anemia. In turn, anemia is often linked to a diet that lacks or is seriously deficient in meat. If the Olduvai child was still being breastfed by a mother whose milk was iron-poor due to a paucity of meat in her diet, then this might explain the occurrence of porotic hyperostosis on the child's cranial bones and his presumed fatal anemia. Alternatively, if the child was already weaned or was being weaned at the time of his death, perhaps it was he who was unable to secure enough meat to eat, leading to his sickness and its osteological manifestation as porotic hyperostosis. Domínguez-Rodrigo and his colleagues go on to speculate that

> [Porotic] hyperostosis is virtually absent in [chimpanzees], our closet living relatives. In contrast, the pathology's relative prevalence in prehistoric hominins [including others than just the Olduvai child] seems to indicate that a significant derivation in hominin metabolic physiology from the ancestral condition occurred sometime after the late Miocene [between 8 and 4 million years ago] split between the hominin and [chimpanzee] lineages. The presence of porotic hyperostosis on the [Olduvai child] indicates indirectly that by at least the early Pleistocene meat had become so essential to proper hominin functioning that its paucity or lack led to deleterious pathological conditions.

Actualistic observations indicate that to have depended only on scavenging in order to secure meat would have been a losing strategy for hominins who were physiologically dependent on animal flesh and who lived in the type of savanna habitat reconstructed for the Olduvai child and other hominins of the same time and geographic distribution. Independent studies of carcass completeness and persistence in savannas by Blumenschine and by Domínguez-Rodrigo both showed that passively scavengeable meat is a rare, unreliable source of sustenance in this kind of environment.

But forget this surprising instance of agreement between the antagonists Blumenschine and Domínguez-Rodrigo, and forget all the previously enumerated failed tests of the "passive scavenging hypothesis"—doubts about the hunting prowess of early *Homo* persist. Again, it is a general incredulity about the physical and cultural competence of early hominins that is the backdrop against which have unfolded the specific objections of their ability to acquire carcasses and their propensity for meat eating. And, even though the discovery of the Nariokotome Boy has helped erode some of the skepticism about the physical capabilities of at least *Homo erectus*, the archaeological record continues to be an equivocal source of support for those who, in contrast, view our remote ancestors as competent predators and vivacious carnivores.

## SLINGING AND SHIVVING

There is no evidence that *Homo erectus* or earlier hominin species produced stone-tipped spears or arrows. Until such time as that changes the only plausible potential weapons of early hominins were rocks, clubs, and pointed sticks, the last perhaps serving as rudimentary, untipped spears. In addition to their defensive and offensive uses against other hominins, our ancestors might have likely thrust or thrown spears at prey and at competitors for carcasses. Although the first known wooden spears are only about 400,000 years old (from Schöningen, a Paleolithic site in Germany), earlier evidence of woodworking by hominins—and, by extrapolation, possible spear production—is found even deeper in antiquity, associated with the initial stages of the evolution of *Homo erectus*. In the late 1970s, archaeologists Larry Keeley and Nicholas Toth studied the microscopic wear patterns on crude stone tools recovered from 1.6 million-year-old archaeological sites at Koobi Fora. Based on comparing use-wear patterns on the Koobi Fora artifacts to those on stone tools that they had replicated and used for a variety of basic activities (butchery, plant food processing, and so forth), Keeley and Toth inferred that some of the worn, polished areas on the Koobi Fora artifacts are best attributed to woodworking activities by their early hominin users.

More recently, an analysis of tear drop–shaped stone handaxes from Peninj also reveals a woodworking tradition at this Tanzanian site nearly 1.5 million years ago (figure 8). Archaeologists are fortunate that phytoliths, robust microscopic particles of silica derived from plant cells, often survive the rigors of fossilization—adhering to artifacts against which plants were worked—when other parts of the plant have long since decomposed. Doubly auspicious, Manuel Domínguez-Rodrigo and his colleagues observed, and were able to classify, phytoliths on several of the Peninj handaxes. Based on their shapes and sizes, the phytoliths are probably derived from acacia trees. Because the Peninj handaxes are large, heavy-duty tools, and because acacias are extremely sturdy trees, the causal link, established by phytolith evidence, between the tools and the trees suggests that the handaxes were probably used to chop acacias. It is easy to conjure lots of reasons why an early hominin might chop down and chop apart an acacia tree and/or its limbs: for firewood, shelter, or bridgework. In addition, Domínguez-Rodrigo and his colleagues suggest spear production as another hypothesis to explain the Peninj phytolith data. Of course, the *possibility* of Pleistocene armaments is one thing, but realistically imagining their mobilization is quite another. A return to the modern world is the best source to model ancient spear use and its potential payoff.

Critics disparage that using modern human behavior as a baseline to investigate *Homo erectus* foraging behavior submits to the "tyranny

**FIGURE 8.** A 1.5 million-year-old stone handax (about 8 inches long) from the Tanzanian site of Peninj. The white dot shows the approximate location where acacia tree phytoliths were recovered. The presence of these phytoliths indicates that early hominins used the tool to work wood (maybe even to make spears). The rendering on the right shows the acacia phytoliths at extremely high magnification. (Photograph courtesy of Manuel Domínguez-Rodrigo)

of ethnography" and obfuscates past hominin uniqueness. This scorn is, however, hurled against a straw man. No sober-thinking scientist suggests that we can uncritically apply the abilities of modern people—with their large brains and sophisticated technologies—directly across millions of years into the past and onto our relatively pea-brained and culturally impoverished ancestors. Instead, a majority of researchers simply recognizes the pragmatism and scientific dividends of an actualistic approach to disentangling the complexities of the past. Additionally, a complementary way some paleoanthropologists use the modern world to understand the past is to establish a frame of plausibility that can serve to bound the behavioral potential of early hominins. Obviously, modern people do (or can do) some things our ancestors did not (or could not). Equally, just because modern people do not engage in certain behaviors does not mean that the same held true for our ancestors. Keeping these simple but essential caveats at the forefront of our deliberations frees us to reject the cynical misrepresentations of actualism as naïve and fruitless and instead allows the present to be productively mined in order to help elucidate the past.

Modern hunter-gatherers use a host of techniques and technologies to acquire meat. They trap small mammals with nets, use rifles to blast away from horseback at large antelopes, and literally walk other animals to death across burning desert sands. Recognizing this rich behavioral variability, the quintessential image of a traditional human hunter is still with a bow and arrow. The bow and arrow is a marvelous mid- to long-distance hunting weapon, and its predecessor, the atlatl (or spear thrower), likewise represents an increase in effective killing range over that of a simple spear. Atlatls and bows function to increase the velocity of the darts and arrows that they project well beyond speeds that can be achieved by unassisted human throwing. The atlatl— usually made of a short length of wood with a spur at one end into which the non-tip end of a dart is inserted to be launched—accomplishes its power by artificially extending the length of the thrower's arm, thus increasing leverage; the bow, by acting as a spring that stores energy in its limbs when its string is drawn, and then transferring that energy instantaneously to an inserted arrow upon release of the string. These high-velocity events mean greater range for the discharged projectile and increased impact into a target.

Indeed, it seems for many anthropologists that the remarkable efficiency of modern atlatls and bows nullifies any consideration that our early Pleistocene ancestors used primitive hand-held spears as hunting weapons. The knock on hand-held spears for hunting by early *Homo* goes like this: spears without stone, bone, or metal tips are incapable of inflicting mortal wounds when cast from mid- to long ranges—but, not only is it difficult to get close to large ungulates, getting close to them is also unbelievably dangerous. Thus, there was no hominin spear hunting in the early Pleistocene; the technology was not there, and without that technology, it was way too dangerous an activity to undertake.

Obviously, this is a topic that cries out for much deeper consideration and more real-world research. To begin that process a digression on that most intriguing and enigmatic of our extinct hominin cousins, the Neandertals, is instructive. *Homo neanderthalensis* is the non-modern hominin species longest known to paleoanthropology. Neandertal remains were first discovered in Engis, Belgium, in 1829 and at Forbes' Quarry, Gibraltar, in 1848 but were unclassified as such until long after 1856, the year that Neandertal bones were found in Germany's Neander Valley (*Neander Tal,* in German). Until recently, even this protracted academic familiarity with Neandertals had done little to clarify their genetic relationship to *Homo sapiens.* They have been considered a subspecies of modern man, and, at other times, have been relegated as an atavistic mistake of nature, having nothing to do with contemporary people. Similarly, headway toward scientifically informed reconstructions of Neandertal foraging capabilities—Were they efficient Ice Age superpredators or shambling scavengers of carrion?—has been characterized by lurching progress, punctuated by damning setbacks.

Recent advances in the laboratory—extracting and analyzing genetic material from Neandertal bones—has gone a long way in clarifying the first matter. Paleogenetic researcher Svante Pääbo's extraction of DNA from a Neandertal fossil initiated a herculean effort that, ten years later in 2009, finally produced the Neandertal genome (although only about 60 percent of it has been analyzed so far). Evidence emerging from the fluorescence of these genetic investigations suggests that Neandertals and modern humans represent distinct species who separated some 450,000–300,000 years ago, and who subsequently interbred when they overlapped in parts of western Europe, for perhaps 15,000 years before Neandertals went extinct around 25,000 years ago.

More relevant to our consideration of Pleistocene spear hunting, recent research on Neandertal foraging has been particularly inventive in its use of actualistic methods. The stereotype of Neandertals as brutish lugs is not without foundation. Neandertal bones, recovered throughout Eurasia, are typically quite stout, with large, raised scars where thick muscles would have attached; analysis of this massive anatomy reveals a species composed of short, brawny people. Detailed studies of the Neandertal shoulder and humerus (the upper arm bone) followed this general appreciation of Neandertals as exceptionally powerfully built people. Studies show that the humeral heads—the ball joint at the top of the humerus that fits into the shoulder blade to form part of the shoulder girdle—of modern athletes who engage in habitual throwing face more backward than the humeral heads of other people, and further, that this unusual backward rotation of the humeral head is more prevalent in the throwing arms of throwers than in their non-throwing arms. Compared to most modern people, Neandertals also have increased humeral backward rotation, or retroversion. But, analyses conclude that the marked humeral retroversion of Neandertals probably has less to do with habitual throwing and more to do with their general body form (most Neandertals had deeper chests than do most modern people; relative chest depth, in turn, probably affects the degree of humeral rotation in order that the humerus is properly aligned in the shoulder joint) and inferred higher overall level of activity as compared to modern humans. That the three known Neandertal skeletons with paired humeri all *lack* significant bilateral asymmetry in the angles of their right and left humeral retroversion would seem to corroborate this hypothesis.

The shafts of Neandertal humeri are, however, asymmetrical, with right humeri thicker front to back than are left humeri. But, compared to modern humans from the Upper Paleolithic (the archaeological period in Europe lasting from about 40,000 to about 11,000 years ago), both right *and* left humeri of Neandertals are relatively wide in this plane compared to their thicknesses side to side. Early modern humans, in contrast, have nicely rounded humerus shafts in cross section. The distinct modern human morphology, among other functions, acts to resist torsion that accompanies a throwing motion. Hence, some researchers argue that by the time of the Upper Paleolithic modern humans were throwing spears habitually.

Also, Upper Paleolithic tools of *Homo sapiens* match, in many ways, the projectiles of contemporary hunter-gatherers. Archaeological assemblages of Upper Paleolithic humans are rife with beautifully executed stone points, which would have been hafted into throwing spears (the wood shafts of which are now disintegrated with time), and atlatls are known from at least 17,500 years ago. The technological similarities between Upper Paleolithic humans and modern people seem to confirm the independent conclusion, based on studies of shoulder and arm anatomy, that early modern humans threw spears.

A look at Neandertal artifacts reveals a different picture. Some types of stone tools they produced would surely have been hafted as armatures in wooden or bone shafts, but compared to samples of Upper Paleolithic artifacts, fewer of them show morphologies and impact fractures indicating that they were deployed as projectiles. In other words, Neandertal artifacts comment only equivocally about whether they threw spears regularly. But, combine that unsatisfying conclusion with the anatomical interpretation that the thick front-to-back humerus shafts of Neandertals functioned to resist "large bending moments engendered by *thrusting* spear use," and the picture of Neandertals as physically rugged, close-killing, big-game hunters was supposedly confirmed with science.

However, it was after paleoanthropologists Thomas Berger and Erik Trinkaus took actualism to the rodeo that the narrative of Neandertals as kick-ass bison assailants was dramatically resurrected in the public imagination. Scientists had long noted fractures and other serious trauma on many Neandertal bones. Suspicious that these injuries might have resulted from the intimate contact with large dangerous prey that is supposedly required by spear thrusting, Berger and Trinkaus sought to confirm this suspicion by comparing damage on Neandertal skeletons to that of modern people who also interact regularly, closely, and physically with imposing hoofed beasts. The frequency and patterning of Neandertal bone breakage, clustered in the upper body, matches closely that seen in rodeo athletes. Berger and Trinkaus were, however, careful to qualify their 1995 "rodeo rider hypothesis." They stressed that the abundance of Neandertal upper body injuries—upon which the comparison to modern rodeo athletes is based—is, of course, *proportional* to the dearth of Neandertal leg bones with injuries. Considering this fact, Berger and Trinkaus suggested that the need for hunting-and-gathering Neandertals to be mobile was as likely a reason as was close-range hunting for the disproportionately low numbers of injured Neandertal leg bones. They wrote, "Those [Neandertals] no longer capable of keeping up with the social group, whether as a result of age or serious lower limb trauma, may have simply been left behind, to die in localities where their remains were not preserved and recovered." In contrast, individuals with superficial injuries to their heads and broken arms and ribs could keep on trudging along with their groups as they moved around the landscape. Fossil evidence accumulated since 1995 shows a pattern similar to that seen in Neadertals—a relative profusion of upper body (including head) injuries to a much reduced frequency of lower body injuries—in more recent prehistoric populations of *Homo sapiens* from the European Upper Paleolithic. In addition, the upper body injuries of some well-known Neandertal fossils are quite clearly the result of interpersonal violence, rather than having been inflicted by counterattack from prey. In light of these recent observations and others, Trinkaus revisited the "rodeo rider hypothesis" in 2012 and concluded that, "[Several] non-mutually exclusive factors combine to create the current pattern. The Late Pleistocene anatomical pattern of trauma likely had multiple causes, and it may well be more profitable to assess each case on its own terms rather than looking for global unifying causes." The important point here, however, is that the caveats of Berger and Trinkaus have had, at least as of this writing, very little impact on the thinking of scientists and the public since the initial publication of the "rodeo rider hypothesis." Evidence of this disregard is seen over and over again in the recent works of paleoanthropologists, science writers, and the producers of television specials, in which they collectively abet the revival of Neandertals as the pugilistic, rough-and-tumble cavemen of days long past. No degree of qualification, from Trinkaus, or anyone else, has been able to shake the picture since.

A problem for enthusiasts of "Neanderthugs" is that, in addition to its formulator applying the brakes on the "rodeo rider hypothesis," there is also an archaeological record that speaks to the idea. As noted previously, wooden spears are known from a site in Germany called Schöningen, which dates to 400,000 years ago, well before the appearance of "classic" Neandertals around 130,000 years ago (figure 9). Further, the three Schöningen spears are very similar to the

FIGURE 9. One of the world's oldest known spears (400,000 years old), shown as it was being excavated from the site of Schöningen, in Germany. The spear is surrounded by bones of animals killed and butchered by early hominin hunters. (Photograph courtesy of Nicholas Conard)

aerodynamic javelins hurled by modern track athletes, in that each of the ancient spears has its center of mass positioned toward the front of its shaft (the Schöningen spears range in length from just under 6 feet to just over 8 feet), near its intentionally tapered business end. The obvious explanation for the congruence between modern javelins and the Schöningen spears is that the latter, like the former, were designed—and designed well—to be thrown. Some analysts wish, however, to evade this judicious application of Occam's Razor and instead situate their interpretation of the Schöningen spears alongside the conclusion that two other spears of the European Paleolithic are probably lances (thrusting spears) rather than javelins (throwing spears). It is true that the Schöningen spears, as well as a spear found among the rib fragments of an extinct elephant in Lehringen, Germany, and a spear point fragment from the English site of Clacton-on-Sea, are thicker and heavier than are javelins throw by modern hunter-gatherers. However, all strength estimates for premodern hominins conclude that they were stronger than us. Thus, ancient humans could presumably have also hurled heavier spears much more effectively than can we. Further, the Clacton-on-Sea "spear" fragment might not even be a spear; alternative interpretations of the artifact are that it was a digging stick or possibly even a snow probe! And, to their great credit, even some of those who doubt that the Schöningen spears were used as javelins still confess that "with respect to hand-delivered spears, the distinction between thrusting and throwing spears is largely artificial: recent (historically known and extant) hunter-gatherers use hand-held spears in both manners."

That admission lies at the heart of a rebuttal we can construct against the general reservation of many experts that there was *any* spear use at all before Schöningen. Again, the (at least) tacit assumption is that substantial distance between an early hominin predator and its prey was necessary for safe hunting. The deleterious injuries supposedly associated with the stereotyped—but, quite possibly erroneous—Neandertal all-in, wrecking ball approach to hunting would seem to substantiate this perception. But, in reality, compared to atlatls or bows—two simple projectile delivery systems with effective killing ranges of 50 yards or more—*any kind of hand-held spear* is essentially a close-range weapon. Steve Churchill, an expert on ancient spear hunting, and his various collaborators, stress this important point repeatedly, including in their wry answer to an old anthropological question that it "is not when spear throwing first occurred in human evolution (as this behaviour undoubtedly dates to the first spears!), but rather when human hunters shifted from a focus on close-range predation with hand-held weapons to a focus on long-range hunting with projectile technology."

Indeed, compared to modern firearms, *even bows and arrows* are relatively short-range killing weapons. Today, many traditional bow hunters treat their arrow tips with poison (figure 10). This not only allows the hunters to successfully pursue extremely large prey animals; it also mitigates the acute rarity of the dropped-it-dead-in-its-tracks "kill shot." But, even with the advantage of poison arrows, and even though maximum shots of anywhere from 50 to 150 yards are obtainable with a simple long bow, most traditional bow hunters, like the Hadza, don't usually bother even to take a shot if they are more than 25 or 30 yards away from their prey—beyond which accuracy declines noticeably. And, it is quite telling that competitive archers, using sophisticated compound bows and high-performance arrows, are, like traditional bow hunters, most accurate at maximum distances of only 10 to 20 yards from target.

Thus, the conversation I initiate in the following pages is *not* predicated on the false choice of short-range spear thrusting versus long-range spear throwing. Instead, the idea I forward is based on the

**FIGURE 10.** A Hadza hunter-gatherer from the vicinity of Lake Eyasi, in northern Tanzania, shooting a poison arrow at an antelope in the distance. (Image courtesy of Henry T. Bunn)

assumption that deploying a hand-held spear at a prey animal—whether thrusting it or throwing it—is *always* a relatively close-range proposition. Thus, the question becomes, What tactic(s) might hominins have plausibly employed to allay some of the risk involved in this type of relatively close-range hunting?, remembering all the while that the thrusting hand-held spear itself provides at least a modicum of distance between predator and prey. The simple stabbing spears of the Fongoli chimpanzees (chapter 3), deployed to hunt minimally dangerous bushbabies, clearly demonstrate this fundamental principle.

## MAN THE AMBUSH PREDATOR

Anthropologists have, at various times and in various ways, tested the stopping power of spears, but none has yet determined the distance at which an untipped spear penetrates the skin, hair, musculature and bones of prey less effectively than does a tipped one. So many variables enter the comparative scheme that they seem impossible to control. Nonetheless, it is still safe to reject the naïve assumption that a stone-tipped spear is, in every case, superior to a spear that terminates in a simple sharpened wooden end. Stones from which projectile tips are made are often flawed by invisible pockets of crystallization or internal fractures, which weaken the points enough that they can shatter harmlessly upon impact with the hide of a prey animal. In one experimental study, stone-tipped arrows shot at 10 to 12 yards away, and stone-tipped spears hurled from 4 to 5 yards away, on occasion, bounced impotently off their relatively thin-skinned dog carcass targets (let me preempt any frothing inquiries from PETA; the dogs were euthanized by a veterinarian and supplied to experimenting archaeologists—neither of whom was me!). Conversely, other researchers have been successful in breaking open the incredibly thick skin of dead African elephants using thrusted stone-tipped spears and atlatl-launched stone-tipped darts.

As for untipped, hand-delivered spears, nonsystematic experiments show that hides of zebra-sized ungulates can be pierced effectively by throws made up to about 15 yards away. That, of course, is with modern human throwers. And, although we have reason to infer (based on their extrapolated large muscular body weights) that our early ancestors were stronger than us, this is an inference that needs to be qualified. Recall from the previous chapter, Alan Walker's hypothesis that, when compared to other apes, humans ultimately conceded powerful for better-controlled muscles. Because bones respond to the work engaged in by their overlying musculature, determining when this concession of power happened in human evolution is, to some extent, possible by measuring the relative robusticity of long bones of various hominin species sampled from different time periods. While *Australopithecus* and the very earliest representatives of the genus *Homo* have very strong, apelike long bones, *Homo erectus* fossils show some reduction in bone strength. Does this mean, in turn, that *Homo erectus* had begun the evolutionary shift toward a modern humanlike musculature, well adapted for the fine motor control needed to make and use stone tools, for running bipedally, and for *accurate throwing?*

Maybe, but other morphological evidence suggests that the throwing ability of *Homo erectus* might still have been poorer than that of fully modern humans. Remember that the humeri of Upper Paleolithic *Homo sapiens* are exquisitely rounded in cross section—molded quite possibly into that form by habitual spear throwing. The arm bones of non-modern hominins do not show this evocative morphology; and their shoulder anatomy also bodes unlikely for regular throwing. The sample of *Homo erectus* shoulder bones is small, some of the best preserved belonging to the Nariokotome Boy skeleton. The boy's collarbones are short relative to those of modern humans of the same developmental stage and body size, and this shortness may have hindered significantly his throwing abilities. Because a long, modern humanlike collarbone—which articulates with the shoulder blade to form a girdle in which the humerus head resides—"forces" the shoulder blade closer to the spine than does a relatively short collarbone, individuals with longer collarbones "have dramatically increased the range of [arm] motion, particularly in a [backward] direction. . . . One potential selective force favoring such an increase in shoulder mobility is throwing, which entails a significant component of [backward] motion of the [extended] arm during the cocking phase." The implication, from paleoanthropologist Susan Larson, is that *Homo erectus*, as characterized by Nariokotome Boy, was a weak thrower.

Clearly, more fossils are needed to verify this argument: Nariokotome Boy was just one of many *Homo erectus* individuals to have lived in the past, and he could have been afflicted with a pathology known in modern humans as short clavicle (collarbone) syndrome. It does seem more likely, though, that Nariokotome Boy represents the typical morphology of *Homo erectus*, especially considering that three relatively new *Homo erectus* clavicles from the 1.8 million-year-old site of Dmanisi, in the Republic of Georgia, display the same anatomy as do the boy's collarbones. Even so, there is no reason to argue that because *Homo erectus* had a non-modern clavicle then it was also necessarily incapable of hunting with spears—especially in light of the notion of close-range thrusting and short- or medium-distance casting. Take, for example, chimpanzees: they obviously have ape shoulders and arms, but they can still throw rocks and sticks across short distances sufficiently well to drive away prowling leopards and attacking lions. And, of course, it is possible to hunt using spears *without throwing* them. Beyond thrusting them, spears can be braced as pikes in the base of a pit-trap (admittedly, a very cognitively sophisticated application of spear technology, for which there is currently no early archaeological evidence).

The potential effectiveness of an untipped wooden spear can be enhanced by heating, and thereby hardening and strengthening, its sharpened business-end. Strengthening a spear's wooden tip means amplifying its power upon impact. And, although the archaeological record is silent on pit trap technology, it is much more tantalizing with regard to evidence of early fire. There are several claims of hominin-controlled fire from East and South African sites that are dated between 1.5 and 1.0 million years old, within the known time range of *Homo erectus*. FxJj 20 East, an archaeological site at Koobi Fora, in northern Kenya, preserves circumscribed patches of heated sediment (similar in size to the circumferences of small, short-term campfires made by modern hunter-gatherers), which are claimed evidence of fire tending by early hominins. Likewise, ancient clumps of burned clay associated with simple stone artifacts at the site of Chesowanja might also indicate the early Pleistocene domestication of fire in Kenya. In South Africa, Bob Brain excavated nearly three hundred burned ungulate fossils from a part of the Swartkrans site that is roughly 500,000 years younger than are the FxJj 20 East and Chesowanja sites. The burned bones from Swartkrans are associated spatially with stone cutting and pounding tools and with an abundance of butchered antelope fossils. Most recently, research at Wonderwerk Cave, also in South Africa, demonstrates the presence of miniscule remains of ashed plants and tiny fragments of burned bones in the sediments of a one million-year-old archaeological level, which has also yielded stone handaxes. The microscopic indications of controlled fire in this ancient level of Wonderwerk are bolstered by the additional discoveries of larger pieces of charred and calcined bones and of stone artifacts fractured in ways characteristic of having undergone heating in fires.

However, each of these claims for the prehistoric domestication of fire is contested on various grounds, which means that the origins of this essential human technology remain an open archaeological question. And, even if the evidence from one or more of these (and/or other) sites is someday verified as genuine, it will probably have little direct bearing on hypotheses of early hominin spear hunting. Their ability to control fire does not necessarily mean, in turn, that our ancient ancestors also possessed spears and heated those spears in the fires that they made; fire is equally useful, if not more useful, for illuminating darkness, for warding off predators, and for cooking food—all purposes that feature much more prominently in most models of fire domestication than does heat treatment of spears. More simply, though, critics of spear hunting can always, in every case, assume their tried-and-true fallback position: that the closeness to prey required for thrusting or short-distance hurling of a spear disqualifies, de facto, spear hunting as an option for premodern hominins.

It is true that proximity to large, wild animals can *sometimes* be perilous. Opponents of early spear hunting point to Berger and Trinkaus's Neandertal study (even though, as discussed previously, it is increasingly disavowed by Trinkaus) and argue that injuries similar to those incurred by modern rodeo performers—an occupation that more than any other places its practitioners in intimate, dynamic contact with large animals—would have been fatal for early hominins, who obviously lacked access to medical care.

There are, however, safer (and smarter) ways to get close to large, gregarious ungulates, thereby minimizing injurious risk to a hunter. One of those ways is by slow, steady advance on foot—a technique that, when used for hunting, is called "approach hunting." Small, solitary antelopes tend to be anxious little creatures, ever vigilant, never quite relaxed. Their natural skittishness makes them more difficult to approach closely than it is to (cautiously) approach larger, social antelopes.

"Ambush hunting" is another technique that puts modern human hunters in close but relatively safe proximity to large game. Ambush hunting requires stealth but not much more than rudimentary observational skills in order to find a game trail or waterhole and set up, in concealment, beside it. Henry Bunn and I have recently proposed one variant of ambush hunting—hunting from trees—as a likely tactic for a hunter with the obvious climbing abilities of our primate forebears, including *Homo erectus*. Modern humans still employ their inherited tree-scaling proficiency when they hunt from arboreal shooting platforms, the tree-stands familiar to any red-blooded American deer hunter. Because they might be attacked from one of many different directions, diminutive prey animals, like rabbits and squirrels, tend to be hypervigilant; a large raptor diving from above is as likely a predator of these small animals as is a coyote coursing through the underbrush. Medium- and large-sized adult ungulates, on the other hand, are most often attacked by predators operating from ground level, so it is unsurprising that the focus of their watchfulness is at that level.

Steve Churchill's meta-analysis of the hunting techniques of modern foragers reveals that they usually employ hand-held spears only when their prey is disadvantaged. For Churchill, a disadvantaged animal is one that has succumbed to any predatory technique that limits its escape or one that has placed itself in a situation that gives its predator the time and opportunity to employ his weapon. For examples, a rhinoceros driven against a game fence by a Land Rover carrying some pitiable trophy seeker on a canned hunt is disadvantaged by circumstances beyond its control, while an antelope feeding dully at the terminus of a deep, steep-sided gulch has disadvantaged itself. Applying Churchill's scheme to hunting from a tree stand shows it to be a predatory tactic classifiable as both ambush (the hunter is concealed from his prey) and also, because the prey does not naturally direct its gaze overhead, disadvantaging.

Churchill's ethnographic survey also makes it clear that approach hunting with hand-held spears is less common than is using hand-held spears in ambush hunts. Only Australian Tiwis and native Tasmanians, of nearly one hundred foraging groups surveyed by Churchill, used hand-held spears in approach hunting. The only other times that hunter-gatherers used hand-held spears was in pursuit hunting, which involves chasing an animal to exhaustion and then dispatching it safely with the spear (this technique often involves the use of horses or dogs as hunting partners), or, in ambush hunting. Based on the framework of plausibility that these modern data provide, it seems most likely that early hominins would have used hand-held spears to hunt primarily by ambush, rather than by other techniques.

Ambush hunting encompasses a wide range of techniques—from a single guy hiding behind a rock, waiting for just the right shot, to a large group of people coordinating their efforts in order to drive a herd of ungulates through the firing path of their concealed compatriots. Along this tactical continuum, hunting from a tree-stand is fairly simple, but it still conveys many benefits to the hunter. In addition to the disadvantaging nature of hunting from above (again, ungulates do not typically look up when scanning for predators), attacking an animal from above also takes the hunter out of potentially harmful physical contact with the prey. Moreover, even though the effective lethal range of a hand-casted spear is unimpressive (for a tipped spear, about 8 yards), hunters can surmount this drawback by constructing an arboreal casting platform that, while still above the heads of (and thus invisible to) their prey, is nevertheless fairly lowly placed—situating the hunter well within the maximum killing range from his target.

Viewed from safely overhead, a quadrupedal prey animal presents a remarkably broad, extremely vulnerable target to the hunter. A shot launched from directly above the animal can incapacitate it in one of at least two ways. A good overhead shot might sever the animal's spinal cord. With the prey so paralyzed, it is also rendered harmless to the hunter when he descends from his killing perch and dispatches the animal at close range. More dramatically, a good (or lucky) hunter can—like a first-rate *torero* thrusting his *estoque* over the bull's head and between its shoulder blades—place an overhead shot so precisely that it ruptures the prey's aorta, a nearly instantaneously fatal wound. And, if instead, the prey is not right underneath the hunter, it serves only to present the animal as an even more expansive target than if it was viewed from directly overhead. In profile, the animal's lungs, diaphragm, and heart are exposed behind and under the shoulder blade, open to absorbing a mortal broad shot, cast on an oblique trajectory from above.

In addition, when a hunter hurls his spear from ground level, in an arc at a target roughly horizontal to him, the weapon's force is diminished across the distance it is cast. If the spear finds its target, it must then confront consecutive layers of thick hair, elastic hide, dense musculature, and tough bones, all of which serve to protect the prey's vital internal organs. A spear's impact with this organic composite of body armor only exacerbates the reduction of its force and, concomitantly, its lethality, as it seeks its target's deeply seated mortality. In contrast, the hunter can mitigate degradation of force by casting his spear from a tree-stand, powerfully, in a downward direction and over a short distance. Because arboreal casting increases the potential force on impact, an untipped spear probably has an important advantage over a stone-tipped one when hunting from a tree-stand. Reviewing the ethnographic use of tipped versus untipped spears, anthropologist

Christopher Ellis concluded that the "brittleness [of stone spear heads] . . . have an effect or influence on the contexts in which stone-headed thrusting spears were employed by limiting use to situations where the danger to the hunter or warrior was constrained." This conclusion speaks to the unreliability of stone points and their potential to fail upon forceful impact, which could leave the hunter (or warrior) in a very precarious position vis-à-vis its prey (or enemy warrior). A spear made of a single wooden stave with a sharpened end is less apt to shatter or snap upon impact. It can also be easily withdrawn from a wound and then stabbed repeatedly into an incapacitated animal, until the coup de grâce.

Of course, just because logic and ethnographic data support the idea of early hominin ambush hunting does not necessarily mean that early hominins *actually* hunted by ambush. But, if one accepts that the archaeological evidence of butchery discussed previously (that is, abundant defleshing and demarrowing damage on meat-bearing bones) indicates that Pleistocene hominins enjoyed early access to large ungulate carcasses, then it is incumbent upon us to explore ways in which they achieved that early access. Aggressive scavenging—forcibly appropriating a carcass from its predator—is certainly the more dangerous of the two major ways that meat eaters can gain early access to carcasses. In fact, confronting a feeding predator is at least as dangerous as (and, in most cases, probably more dangerous than) coming into close contact with large herbivorous prey. A grass eater's food is ubiquitous; rather than confronting a predator, a good, evolutionarily fit strategy for an antelope is to simply move out of harm's way and on to a different pasture. In contrast, a predator's edible prize is usually hard won, which means it is likely to put up at least some fight to protect it from a pirate.

Hunting is the other main way that foragers can gain regular early access to carcasses—and spear hunting by ambush (and/or approach) seems the most *plausible* current hypothesis of how Pleistocene hominins hunted. That is not to say that other *reasonable* hypotheses haven't been offered. One idea that garnered recent headlines (and is, on the face of it, reasonable) is that early *Homo* hunters ran antelopes to exhaustion; once incapacitated, the poor, ragged-out creatures were safely dispatched by hominins at close range. Like spear hunting, pursuit persistence hunting by running has precedent in the modern world. Running persistence hunting is, however, only carried out by foragers living in very hot and very dry, open habitats, such as by some San hunters in the Central Kalahari of Botswana. It is in this type of extreme environment that hunters can most effectively tire their prey by driving it relentlessly across loose sand and denying it any relief of shade and water. The flat, barren ground of the Central Kalahari also allows hunters to successfully track their prey when it, inevitably, breaks visual contact with them. But, even under these most "ideal" ecological conditions for persistence hunting by running, it is still a very rare undertaking, and none too successful compared to other hunting strategies.

An additional problem for those who propose that early *Homo* used running to persistence hunt is that a Kalahari-like ecology, ideal for the tactic's success, does not match the habitats that we understand early *Homo* to have occupied. Instead, early *Homo* lived in savanna-woodlands, with substantial grass and tree cover that would have complicated tracking considerably. In these kinds of more densely vegetated habitats, tracking is a challenge even for fully modern people (for example, Hadza men, hunting in savanna-woodlands, will give up tracking a wounded animal when it is apparent that its spoor is lost and other, more reliable—if not more profitable—foraging opportunities are available). Add to this the fact discussed previously that running hunts are only rarely successful even when undertaken by modern foragers in the hot, dry, open habitats in which greatest success is predicted. Indeed, when people hunt by persistence, it is usually by walking—*not running*—animals to exhaustion. Most prey animals can easily outrun their human hunters before they are incapacitated by fatigue, so expert tracking is essential to successful persistence hunting. But it is nearly impossible for a human to both run and, at the same time, carefully track an escaping animal.

Supposedly, a recently ballyhooed "study" of a group of Kalahari San substantiates the "endurance running hypothesis" of early hominin hunting. In reality, the paper in question merely reports anecdotal observations of just ten running persistence hunts by these San over the course of *twenty years*. Only two of the hunts were undertaken spontaneously, as the results of decisions that the hunters made on their own; the other eight hunts were prompted by the author of the study so that he could film them for television documentaries. Overall success of the running technique was also equivocal. The two spontaneous hunts *were* successful, but each was undertaken by the same four men; perhaps these four guys were uncommonly good at this particular, rare type of hunting, or maybe they were lucky twice. Further, only three of the eight prompted hunts were successful, *even though those hunts were commenced from an off-road vehicle and by hunters who refilled their water bottles during hunting*—decided non-options for early Pleistocene hominins! Finally, scrutiny of the San data shows that the "average speed of all observed persistence hunts . . . was 1.72 [meters per second], a speed at which walking is usually preferred." Running very slowly like this is a very energetically inefficient (and ridiculous) way to run. It makes no sense, so why not just instead walk—without your head bobbing up and down—so that you are able to track your quarry more efficiently?

Together, these caveats diminish the *plausibility* of a hypothesis proposing that early hominin hunters regularly combined running and tracking as a successful hunting tactic in their savanna-woodland habitats of nearly 2 million years ago. In contrast, the hypothesis of early *Homo* ambush (or possibly approach) spear hunting is borne out by a recent study of ungulate prey from the 1.8 million-year-old FLK Zinj site at Olduvai Gorge. By wide consensus, FLK Zinj—with its hundreds of cut-marked and hammerstone-broken bones from dozens of large mammals—provides the very best early Pleistocene evidence of systematic butchery and meat eating by the genus *Homo*. Hominins were, by far, the dominant, if not exclusive, accumulators of ungulate bones at FLK Zinj. By examining the eruption and wear patterns of the teeth of antelope jawbones from FLK Zinj, Henry Bunn and I were able to place individual animals into age categories (as in all mammals, antelope teeth emerge and are ground down at, respectively, known ages and predictable rates, allowing us to estimate age from jawbones with teeth). We then showed that the distribution of the ages at death of those animals does *not* meet predictions of hypotheses that the Olduvai hominins either passively scavenged from kills of large cats or hunted by running prey to exhaustion.

Most of the small antelopes (with estimated live body weights of less than about 150 pounds) represented in the FLK Zinj bone assemblage were old males when they died. This distribution of age at death is completely different than is the mortality pattern created by modern leopards when they kill and eat small antelopes, which is instead dominated by antelopes of prime adult age. Using this actualistic knowledge to interpret the formation of the FLK Zinj bone assemblage requires that we reject the hypothesis that early *Homo* scavenged passively from the carcasses of small antelopes, which were killed, dragged into trees, and then abandoned by leopards after their partial consumption. Admittedly, the predominance of old, small antelopes at FLK Zinj is also expected from hominins running down the older and weaker members of a herd.

But, as discussed previously, the moderately to thickly vegetated habitats (challenging to even brilliant, modern trackers who are taking their time and proceeding carefully) of early *Homo* makes it extremely unlikely that the small antelope bone sample from FLK Zinj was created by hominins through persistence hunting by running. Bunn and I returned to our observations of the Hadza to put the situation in perspective: "Hadza foragers, who are skilled, lifetime trackers, routinely lose arrow-shot, bleeding prey in such [well-vegetated] habitats. And, they would not even consider trying to track healthy animals through heavy vegetation at a jogging pace. Attributing to early *Homo* tracking skills well beyond those of modern foragers is simply unrealistic." Again, the modern world—and an actualistic appreciation of the complexity of it—provides a framework so important for assessing the plausibility of paleoanthropological hypotheses.

Turning to the dental sample of large antelopes (with estimated live weights in excess of 200 pounds), FLK Zinj is dominated by prime-age adults. If, as posited by Rob Blumenschine, early *Homo* was passively scavenging large antelope carcasses only after they were abandoned by lions, then the age distribution of the resulting antelope tooth sample should mirror the living structure of an intact antelope herd. Operating as ambush predators, lions use stealth and explosive power to kill opportunistically and are thus unselective as to the sex and age of their prey. They kill what is available when they are able. However, if early *Homo* was instead running down large antelopes, then the resulting pattern of antelope mortality should be one dominated by very young and very old individuals—the weakest, most vulnerable members of the prey population. In modern settings, it is these most defenseless members of a population that are targeted by hyenas and other predators that run down their prey. Thus, both the "passive scavenging" and the "endurance running-hunting hypotheses" fail completely to predict the *actual* mortality pattern of large antelopes in the FLK Zinj bone assemblage—one that is dominated by the teeth of prime-aged adults. By the process of elimination, and the application of the principle of plausibility (that is, considering the probable cognitive and physical capabilities of early *Homo* compared to modern humans; assessing reconstructed habitat parameters; and making an educated guess about probable basic hunting technologies of the time), Bunn and I therefore concluded that ambush hunting by early *Homo* was likely the major origin of antelope mortality at FLK Zinj.

Positing that *Homo erectus* spear-hunted by ambush forces us to explore even more deeply the requirements and implications of this tactic. Hunting alone is a successful technique for modern Hadza men. A large part of what makes a Hadza man a successful lone hunter is that, with his bow and poison arrows, he possesses a relatively advanced projectile technology. Remove the advantage of this sophisticated, long-range hunting weaponry and it would likely diminish the man's success. The archaeological evidence of butchery at early *Homo* sites indicates that those early hominins, like modern Hadza hunters, consistently acquired large animal carcasses. However, unlike the Hadza, our Pleistocene ancestors accomplished this feat by using, at best, only close-to moderate-range projectile weapons. Thus, it is reasonable to suggest that another variable allowed *Homo erectus* to be seemingly as success-ful hunters as are cognitively modern, superiorly armed Hadza men. It might have been hunting in groups that bought *Homo erectus* up to par with the standard set by a single Hadza man hunting with his bow and poison arrows. Two intentional consequences of group hunting—in-creased vigilance and defensive recruitment—could have also reduced individual risk in (accidental?) contact with large prey (falling out of a tree? being turned upon during an approach hunt?). But, reconstructing early hominin group size is a Sisyphean task; and deducing the size of a hunting party, a subset of that larger group, would be impossible.

When modern humans and chimpanzees *do* hunt in groups, it is nearly exclusively a male activity. All the extant African apes are philo-patric, with males remaining in their groups of birth. Thus, those stay-at-home males are surrounded by their blood relatives, including very closely related brothers and half-brothers, for their whole lives. It seems likely that this primitive ape trait was also a feature of the most recent common ancestor of chimpanzees and humans, as well as of the earli-est hominin species. The principle of inclusive fitness predicts that this simple aspect of close relatedness would have resulted in dependable cooperation between males within a group: your brother's reproduc-tive success is also a genetic achievement for you. And, observations of extant African apes and humans, allows us, within this general philopat-ric framework of social organization, to focus in even more specifically on group hunting behavior. As discussed in chapter 3, hunter-gatherer men form natural within-group coalitions (as do male chimpanzees) in order that they can successfully eliminate male competitors in neigh-boring groups, expand territory, and improve their access to food and females. It would seem that the bonds forged in those "raiding coalitions" hone a type of male–male cooperation that can be easily transferred to hunting activities. Males come to know and rely on one another. In humans, sensitivity to comembers of a raiding or hunting party is enhanced by sophisticated language abilities—not just vocal communication, but also hand signals and other cryptic, nonverbal exchanges.

To contend that group raiding tendencies are also employed in hunting is *very different* than to simply equate raiding and hunting. Rather, propensities and skills developed and relied upon in one "hunt-and-kill" activity well serve the individual and group in another "hunt-and-kill" activity. With the exception of the common fact that raiding and predation are male group activities, those propensities and skills are expressed quite differently in chimpanzees and humans: chimpanzees (excluding the informatively anomalous population at Fongoli) act largely emotionally and in intimate quarters, neutralizing the objective (enemy or prey) with brute force; humans act cerebrally, most often neutralizing the objective with a weapon and from at least a minimal distance. In both cases, it is appropriate to view the connection between interpersonal violence and predation as application of a circumscribed behavior set to functionally similar *but motivationally unrelated* "problems."

For chimpanzees and humans, killing (or maiming irreparably) another organism is the proximate goal of both raiding and hunting. The ultimate reason for both raiding and hunting is also the same: to increase Darwinian fitness. It is on an important intermediate plane between proximate and ultimate levels that predation and interpersonal violence are most clearly distinguished by the *way* in which each contributes to increased fitness. Most directly, successful predation yields the sustenance (energy, nutrients, and calories) that an individual needs in order to live and, in turn, to pursue opportunities to reproduce. Additionally, good male hunters probably also reap reproductive benefits through resource exchange with females (meat for sex), and, more indirectly, through enhanced prestige in the eyes of females. For male chimpanzees and hunter-gatherer men, inflicting violence on a neighboring group holds the potential to expand territory (and, by extension, gain access to essential resources), eliminate sexual competitors, and provide mating opportunities with new females.

Setting aside whether or not *Homo erectus* operated as a group hunter, it is appropriate to close this chapter with a return to those unusual hunting chimpanzees of Fongoli, who provide essential context for the more encompassing hypothesis that *Homo erectus* was a spear hunter. The Fongoli chimpanzees provide this context, first, by way of comparison: they reveal that general skepticism about spear hunting— that it was beyond the conceptual grasp and technological capabilities of early Pleistocene hominins—also denies *Homo erectus* an apparently quite primal logic, as well as even the most rudimentary skills of manipulation. The Fongoli chimpanzees, with half or less than the cerebral mass of *Homo erectus,* make and use spears effectively on a regular basis. Second, the Fongoli observational data render irrelevant, to a large extent, the open question of whether emotionally checked hunting was simply a component of an overarching, generalized control of emotion that may have characterized *Homo erectus* as a species. The Fongoli chimpanzees demonstrate that emotionally controlled hunting—as facilitated by distance-enhancing weapons— can simply be a *situational* foraging tactic rather than the expression of some immutable instinct; by nature, chimpanzees are much more emotionally driven than would be inferred from witnessing a bushbaby hunt at Fongoli.

General emotional control in hominins may not have yet developed by the time of *Homo erectus*. But, the archaeological record of *Homo erectus* implies strongly that the species applied emotional control, at least situationally, when it hunted, just as do the Fongoli chimpanzees (of course *Homo erectus* tangling with large, potentially dangerous ungulates, operated on an entirely different scale than do Fongoli chimpanzees taking small, relatively harmless bushbabies). The success of emotionally controlled hunting for *Homo erectus* is attested by its behavioral traces, which include proxy archaeological measures like prey body part representation (portions that in life carried large, favored chunks of meat), butchery mark data (defleshing cut marks on those favored portions), and mortality patterns (inconsistent with scavenging passively from the kills of other predators) that show recurrent and consistent access to the best parts of complete carcasses of large ungulates. Regular access to intact carcasses, facilitated by emotional control, would have reinforced and eventually fixed a behavioral complex in *Homo erectus* until decoupling of predation and aggression came to characterize the species as a whole—and tactic was, thus, ultimately transformed into strategy. Modern humans, the direct descendants of *Homo erectus,* follow the same strategy today in order to obtain carcass foods. But, rewind the tape of prehistory back beyond us and beyond *Homo erectus*. What about Raymond Dart's original killer ape, *Australopithecus?*

# Death from Above

EAGLE CARRIES OFF A CHILD
Seized a Little Girl by Her Clothing—
Finally Driven Off
*Special to the New York Times*

WILLIMANTIC, Conn., Sept. 12—All Windham County is
excited over the carrying off a four-year-old child by an eagle
at what is known as Gurleysville, near Mansfield. The child
is said to have been Anna, daughter of Hermann Hertz. She
was playing with companions when the eagle came down and
fastened its claws in her clothing.

The bird succeeded in carrying the child a short distance,
when it alighted, apparently to take a better grip on the
child's dress. A number of older children had then joined the
party. They attacked the eagle and beat it away, rescuing the
child before the bird swooped down on them again. Three of
the boys who fought away the eagle were badly scratched by
its claws.

All the old hunters in the vicinity have organized and are
trying to locate the eagle.

—*New York Times*, September 13, 1899

Watch most any monkey or ape and it's hard not to conclude that
primates are, in general, inherently clever. It is doubtful that *Australopithecus* was disposed any differently. The innate cleverness of ape-men
is, however, largely invisible to us today, millions of years after they
went extinct. In particular, *Australopithecus afarensis* (Lucy's species)
and *Australopithecus africanus* (the Taung Child's species) are derived

from geological deposits that lack any co-occurring traces of material culture, which are physical expressions of cognitive heft; there is no archaeological record of these early forms of ape-man. In contrast, fossils of robust australopithecines are often found in spatial association with stone tools and, in South Africa, with authentic bone tools. But determining that these large-toothed ape-men made the various artifacts that have been excavated along with their bones is complicated by the contemporaneous presence of *Homo erectus* remains at the same ape-man sites (or, in some cases, by at least presence of *Homo erectus* in the same region at the same time).

Simple stone tools—sharp-edged flakes struck from larger river cobbles and the cobbles themselves—are, however, found along with butchered ungulate fossils at the archaeological sites of Gona and Bouri, both in Ethiopia's desolate Afar Depression, both dated around 2.6–2.5 million years old, and both presumably formed by the advanced ape-man species *Australopithecus garhi*. Conventional anthropological thinking—that *Homo* was the first and only tool-bearing member of the hominin lineage—did not predict the coincidence of ape-men and artifacts that is suggested at Gona and Bouri. Remember that the earliest appearance of *Homo*, at roughly 2.3 million years ago (see chapter 2), postdates the Gona and Bouri finds by, respectively, 300,000 and 200,000 years. Of course, there is no way to *definitively* attribute the stone tools and butchered animal bones from Gona and Bouri as the work of *Australopithecus garhi*, but the circumstantial evidence (that is, the absence of early *Homo*) makes a fairly compelling case that *Australopithecus garhi* made and used stone tools hundreds of thousands of years before *Homo* evolved.

*Australopithecus garhi* has several morphological features that might indicate that it was the direct australopithecine ancestor of the genus *Homo*. The cranial anatomy of *Australopithecus garhi* is distinct from that of other contemporaneous ape-men: the size relationship of its canines and molars is like the size relationship of those teeth in early *Homo*; the shape of its premolars is like that of early *Homo;* and the ratio of its upper arm bone to its thigh bone is like that of *Homo*.

Behaviorally, the scant butchery evidence from Gona and Bouri indicates that at least some hominins of 2.5 million years ago (possibly including *Australopithecus garhi*) were, like modern hunter-gatherers, able to obtain and render carcasses of animals substantially larger than themselves. The large size of the prey of modern people compels hunters to make decisions about processing carcasses at the sites of death and about long-distance transport of carcass parts to base camps, decisions that, because of their smaller quarry and transient evening gathering sites, chimpanzees are not forced to make. The large carcasses obtained by human hunters also stimulate complicated sharing of carcass parts among group members, dictated by complex social relationships that are often beyond simple filialness and dominance-based fealty—a state of affairs antithetical to that of chimpanzees, for whom dominance, status, and selfish trade (for example, to solidify relationships with male coalition partners) are the prime determinants in division of meat. It is still an open question whether the large prey discovered in Stone Age archaeological sites means that our ancestors dealt with carcass division, transport, and sharing as do modern humans.

But, as for reconstructing carcass procurement strategies of early hominins, the array of butchered skeletal parts from Gona and Bouri is informative: the fossil assemblages from those sites include ungulate bones that display the same butchery patterns as on bones defleshed by modern human hunter-gatherers and by *Homo erectus*, with cut marks occurring predominantly on the once-meaty shafts of upper leg bones. The bone assemblages from Gona and Bouri are, however, small, and this is why some researchers relegate the sites, contending that early access to carcasses (and, by extension, hunting and/or aggressive scavenging) by hominins was firmly established only much more recently, in the time of *Homo erectus*. An alternative view also worthy of investigation is that it is not that hominin abilities changed between the time of Gona and Bouri and later days, but simply that hominins hunted (and/or aggressively scavenged) with *increased regularity* over time. This alternative hypothesis also accommodates the more robust, more widespread pattern of hominin early access to carcasses that is apparent in the post-Gona and -Bouri stages of the archaeological record. Obviously, more work is needed to clarify this issue.

In the meantime, one of the most interesting aspects of the 2.6–2.5 million-year-old cut mark data from Gona and Bouri is the simple fact that *any kind* of butchery is evident at this, the very beginning of the archaeological record. The coincidence of these two major classes of archaeological data—tools and bones bearing tool-induced butchery marks, which link causally hominin behavior and resource exploitation—seems to corroborate the long-held belief that hominins invented stone tools so that they could use them to reduce animal carcasses into edible portions.

We may have reached the limit of the detectable archaeological record with Gona and Bouri. Geological deposits that predate the sites are common enough in East Africa, and those deposits also contain a copious fossil record that includes hominin bones. But, no stone tools or butchered animal bones have been recognized in contexts even just a few hundreds of thousands of years (a mere instance on a paleoanthropological scale) older than Gona. Prior to *Australopithecus garhi*, ape-men might have relied (as do modern chimpanzees) on perishable wood tools and/or unmodified stone tools to obtain and/or butcher animal carcasses. In fact, a recent high-profile report in the journal *Nature* describes two purportedly butchered ungulate fossils from the 3.4 million-year-old *Australopithecus afarensis* site of Dikika, Ethiopia, some 800,000 years older than is Gona. The study's authors argue that marks on the Dikika bones indicate that they were defleshed and demarrowed by ape-men using unflaked but naturally sharp rocks (it is probably relevant that none of these hypothetical sharp edged rocks were found with the bones at Dikika).

It was not just the popular press, but also a majority of paleoanthropologists who swooned over the bold conclusion that the Dikika fossils are the world's oldest butchered bones. Seeing things differently, Manuel Domínguez-Rodrigo, Henry Bunn, and I were the first (and, as of this writing, only) taphonomists to criticize the Dikika claims in peer-reviewed scientific outlets. Our critique concludes that the published evidence does not, in fact, support the identification of bone surface marks on the two Dikika fossils as unequivocal butchery damage. We further assert that any equivocation surrounding butchery claims of this great antiquity (again, nearly 800,000 years older than the oldest known butchery marks from Gona, where cut-marked animal bones *are derived from fine-grained sediments and in spatial association with sharp-edged, hominin-produced stone tools*) should lead to rejection of such claims. The equivocation stems from two facts: first, the Dikika fossils derive from coarse-grained, potentially abrasive sediments that contain big chunks of sharp particles; and, second, the damage on the surfaces of the Dikika fossils is indistinguishable from that imparted on the surfaces of modern bones that have been trampled on and/or otherwise pushed around randomly in the same kind of coarse-grained sedimentary substrates. In fact, Domínguez-Rodrigo and his colleagues were able to reproduce, by trampling on modern bones lying on coarse-grained surfaces, incidences of damage that match exactly, mark for mark, those observed on the Dikika fossils. Cutting bones with unflaked but naturally sharp rocks might, as the Dikika researchers claim (but as our own experiments using naturally sharp rocks do *not* support), produce marks that look like those on the Dikika fossils. However, that result would in no way nullify the fact that bone surface damage created by random, non-butchery-related processes (like trampling) *also* mimics the Dikika marks, nor would it change the abrasive sedimentary context (ripe for inducing cut-mark mimics) of the fossils.

## GRECO-ROMAN APE-MEN

Even though we explicitly reject the claims of *Australopithecus afarensis* butchery at Dikika, I still think that pre-Gona hominins hunted animals and ate meat. Both modern humans and chimpanzees, our closest living relatives, are competent predators and enthusiastic meat eaters. It seems likely then that meat was also important in the diet of the most recent common ancestor that we shared with chimpanzees, with the importance of carnivory retained in the human and chimpanzee lineages as they each evolved independently after diverging between 8 and 4 million years ago. Further, chimpanzees that live in markedly seasonal savanna-woodlands, like those reconstructed for early hominins, are the populations that hunt the most often and most successfully. This suggests that chimpanzees hunt, ultimately, in order to make up intra-annual nutritional shortfalls when protein and fat are unavailable in other edible resources. If that is the case, and if the model is also applicable to early hominins in similar environments, it strengthens the argument that pre-Gona hominins ate meat regularly but still fails to inform about their exact strategies to obtain meat.

Lack of recognizable stone tools in the paleoanthropological record prior to 2.6 million years ago might indicate that early hominin meat eaters were limited to consuming small animals, which they could process using just their hands and mouths. Large carcasses, with heavier hair cover and tougher, more pliable skin, required either cutting tools or skin that was breeched first by the sharp teeth and tearing claws of an earlier, non-hominin consumer. But actualistic research in savanna-woodland habitats shows that opportunities to scavenge large carcasses are rare and ephemeral there. Thus, a scavenging-only strategy in this kind of habitat is not a viable option for an animal that is counting on meat to offset protein and fat shortfalls when other foods are unavailable or when those foods that *are* available lack these essential nutrients.

The spear-hunting chimpanzees of Fongoli demonstrate that a primate need not be terrestrially fleet footed, or even particularly agile in trees, in order to use rudimentary distance-enhancing weapons to become a *regularly* successful hunter of small game animals. Likewise, sessile infant mammals, concealed but undefended when their mothers left them to forage, might have been targeted by hominin predators. More investigation is clearly needed in this area, but this model of hunting by basal hominins—in which small animals in seasonal savanna habitats were harvested using rudimentary, perishable tools (like sticks) or simply by hand and mouth—is not novel. Louis Leakey, for one, relished his skill at catching small antelopes and springhares (an African rodent) by hand and argued that early hominins would have been just as capable.

As anthropology's archetypal aggregator of hunting and aggression, it is easy to imagine Raymond Dart's reaction to this model. The gauge is found in his response to a different but similarly prosaic claim for *Australopithecus*. In 1981, Bob Brain had just discovered *genuine*, not "osteodontokeratic," bone tools made by early hominins at Swartkrans. He showed the tools to Dart, who

> was over 90 years old at the time and his eyesight was failing, but he felt the smooth, tapering points [of the tools] with his fingers. Then he said: "Brain, I always told you that *Australopithecus* made bone tools, but you never believed me! What were these used for?" I replied that I thought that they had been used for digging in the ground [in order to extract edible roots]. Dart slumped back in his chair with a look of total disbelief on his face. "That," he said, "is the most unromantic explanation I have heard of in my life!" He then grabbed the longest of the bone points and stuck it into my ribs saying, "Brain, I could run you through with this!"

Such was Dart's good-natured "concern." Could his killer ape really have been such a fop?

Recent research on, of all things, leg bone length in *Australopithecus* would argue that ape-men were no pacifists. Like most living, large primates (except people), extinct ape-men had relatively short legs for their body sizes. The contrasting long legs of *Homo* (including even those of its earliest species, like *Homo erectus*) probably made it a more efficient bipedal strider than were the australopithecines. But, the anatomy of ape-man hips, legs, knees, and ankles indicates that its species were also quite capable terrestrial bipeds. So, why didn't natural selection also favor longer legs in the ape-men during their 3 million-plus-year earthly tenure? Were their short legs retained by some counterbalancing selective force acting on another aspect(s) of their biology?

The behavior of modern apes helps address these queries. When used in combination with their relatively long arms, the short legs of apes allow them to climb vertical tree trunks very efficiently. (A telephone lineman, with arms and legs of subequal length, must emulate an ape's anatomy—using a climbing strap to artificially increase the lengths of his arms—in order that he can easily scale an upright telephone pole.) Another advantage of short legs for an ape is that they increase the animal's stability when it stands on all fours in tree branches. It is unclear whether the need for these same behavioral and postural capabilities—vertical climbing and standing on branches—exerted selective pressure in the same way on *Australopithecus* leg length, but many researchers are unconvinced that was the case. Instead, biologist David Carrier has offered yet another explanation for the retention of short legs in the terrestrially bipedal australopithecines.

Many animals, including living apes and extinct ape-men, are or were sexually dimorphic, with obvious physical differences between males and females. Sexual dimorphism is expressed, for example, in different colors of male and female pelage or in the presence of antlers on males and their absence on females. In primates, sexual dimorphism is most commonly revealed in the larger bodies and canine teeth of males versus the smaller bodies and canines of females. As discussed in chapter 2, primate species that display body and canine size sexual dimorphism are ones whose males typically engage in intense competition for opportunities to mate with females. A male who is big and heavily armed enjoys success in this kind of social system.

Ape fighting is all-out, no-holds-barred ferocity. The normally reserved Jane Goodall describes chimpanzee scraps as an orgy of "hitting, kicking, stamping on, dragging, slamming, biting, scratching and grappling." Especially vulnerable (and cherished!) protrusions, like ears and testicles, are often torn off in the course of ape-on-ape brawls. Carrier wades into this carnage with the clinical explication that

> great apes . . . fight with their [arms] from a bipedal stance on the ground. . . . Characters that improve strength and stability in a bipedal stance should enhance fighting performance. Short [legs] lower the center of mass in a bipedal stance, increasing postural stability. Short [leg bones], all else being equal, also increase the horizontal (i.e., shearing) forces that can be applied to the [ground], by reducing the length of the ground reaction force moment arm at the hip joint. . . . Thus, shorter legs may improve the fighting performance of apes.

Carrier goes on to demonstrate that apes with high levels of aggression between males (as indicated, in turn, by high degrees of body and canine size sexual dimorphism) also have relatively short legs.

He concludes, therefore, that short legs in fully bipedal ape-men probably indicate strong selection for regular and elevated levels of combat between males rather than for apelike climbing.

Carrier's hypothesis obviously contradicts Owen Lovejoy's model that the (inferred) humanlike aspects of *Ardipithecus* behavioral biology (especially monogamy and pair bonding, reduced intragroup aggression, and increased cooperation between males; see chapters 2 and 3) were retained and even enhanced in its assumed direct descendant, the genus *Australopithecus*. The contradiction between Carrier and Lovejoy becomes difficult to sustain knowing that, even in spite of their short legs and sexually dimorphic bodies, the difference in heft between males and females of most species of *Australopithecus* was still not as extreme as that of chimpanzees or gorillas, living species with relatively intense competition between males for access to females.

It is true that *Australopithecus robustus*, a fairly recent and anatomically specialized form of ape-man, seems to have had an elevated level of body size sexual dimorphism, but this might be explainable by factors other than aggressive competition between males. As the Pleistocene advanced, the African climate continued to degrade, becoming cooler and drier. In turn, that climatic deterioration caused australopithecine habitat to become less productive and forced ape-men to respond accordingly, by expanding their foraging niches into evermore open, dangerous areas, rife with predators. Tim White and his colleagues suggest that male australopithecines (under Lovejoy's model of early hominin sociality, foraging to not just feed themselves, but also their pair-bonded mates and offspring) might have evolved larger bodies as a response to the intense predator pressure that characterized these new, more hazardous surroundings. If they could not evade the gambits of predators, large-bodied males (foraging in cooperating groups composed mostly of brothers and half-brothers) might well have been better able to protect themselves from attack and launch more effective counterattacks than could smaller males.

## UNHOLY EUCHARIST?

Even if male australopithecines lacked serious, mating-related aggression, that does not necessarily absolve the ape-men of blood sin in toto. Taphonomic observations of an *Australopithecus* fossil from Sterkfontein seem to indicate quite dramatic interpersonal violence among early hominins. Recall that Sterkfontein is the cave site in South Africa where, in 1936, Robert Broom found the first adult specimen of *Australopithecus africanus*. Subsequently, Sterkfontein has proved to be one of the most productive early hominin sites in the world. Most recently, paleoanthropologist Ron Clarke is excavating his discovery of the skull and nearly complete skeleton of an australopithecine dubbed "Little Foot." The tale of Clarke's paleontological sleuthing (begun when he found the ape-man's previously unrecognized foot bones—hence "Little Foot"— in a laboratory storage box marked as containing antelope bones, and then continuing in the very deepest, darkest underground portion of the cave for the rest of the skeleton), and his plucky opposition to the Byzantine forces of a university administration and supposed colleagues allied against him every step of the way, deserves its own book-length treatment. Suffice to report here, Clarke's discovery corroborated an earlier one, which indicated to him that Sterkfontein was home to more than just one species of *Australopithecus*, as is typically conjectured. Clarke has demonstrated that Little Foot's species is different from *Australopithecus africanus* in that jaws of Little Foot's kind not only have large premolars and molars (like all australopithecines), but, unlike most other ape-men species (except *Australopithecus garhi*), lso have large incisors and canines. Little Foot and similar fossils from Sterkfontein and Makapansgat are also unlike *Australopithecus africanus* in having very bulbous cusps on their premolars and molars, in having a flat depression above their brows, as well as a wide braincase, and in having cheek bones that are shifted dramatically forward on the face. Clarke argues that Little Foot and its cohort of like fossils should be called *Australopithecus prometheus* in recognition that the first discoveries of this taxon were made by Raymond Dart in the 1940s at Makapansgat. Dart placed those large-toothed, decidedly non-*africanus* fossils from Makapansgat into the new species *Australopithecus prometheus*—and although the majority of subsequent researchers has disregarded Dart's taxonomic opinion, Clarke has steadily increased the number of fossils from Sterkfontein that, based on their morphology, are not easily placed in *Australopithecus africanus* but that are nonetheless contemporary with that species.

Clarke's work over the years at Sterkfontein has also led to revelations about the more recent end of the time range for South African australopithecines. In 1976, Alun Hughes, the longtime director of research at Sterkfontein, discovered a 2 million-year-old skull designated by the catalog number StW 53. Intriguingly, stone tools were found in the same geological level as StW 53. The spatial association between the hominin fossil and artifacts was used to bolster the opinion that StW 53 represented early *Homo*, because this genus (at least before the discovery of *Australopithecus garhi*, as discussed previously) was assumed to be the first of the hominin stone tool makers and users.

Ever the dissident, Clarke challenged the taxonomic designation of StW 53, arguing that its anatomy is more typical of *Australopithecus* than of early *Homo*. Relevant features of StW 53, highlighted by Clarke, include an *Australopithecus*-like brain size; a braincase that is restricted and narrow in the front, consistent with *Australopithecus africanus*; a flat nose like *Australopithecus* and apes and unlike early *Homo*; large teeth; and premolars and molars that increase in size from the front to back of the mouth, typical of *Australopithecus* but not of *Homo*. In addition, Clarke's careful investigation of the Sterkfontein geology demonstrated that the stone tools found alongside StW 53 are actually not contemporaneous with the skull. Instead, the artifacts had eroded over time into the older StW 53 deposit from a younger, overlying portion of the site. This all means that the proposed functional relationship between the skull and the stone tools is no longer viable.

It does appear, however, that the StW 53 skull is associated in another very intimate way with stone tool technology. During my doctoral research, I observed a series of small striations on the right upper jawbone of StW 53, where this bone rises to articulate with the cheek bone, or zygomatic. Unbeknownst to me, the same observation was made ten years earlier by paleoanthropologists Tim White and Nicholas Toth. But White and Toth held back publishing their results because StW 53 had not yet been fully described.

When we finally became aware of and discussed our independent observations, White, Toth, and I agreed that the linear damage on StW 53, which occurs in three distinct sets, is cut marks inflicted by a sharp stone cutting edge (figure 11). The cut marks form a pattern on the face that butchers often produce when they cut through the masseters (the large chewing muscles that you can feel flex on the sides of your face when you clench your teeth) in order to sever the connections that they serve between the lower jaw and along the cheek bones, as the cheek bones run from the corners of the face to the ears. In addition, the cut marks on StW 53 are tucked up within the hollow of the cheek bone—an area that we argue was naturally protected, unlikely to acquire the types of random scratches imparted by animals trampling around the cave floor and pushing the bone across abrasive sediments.

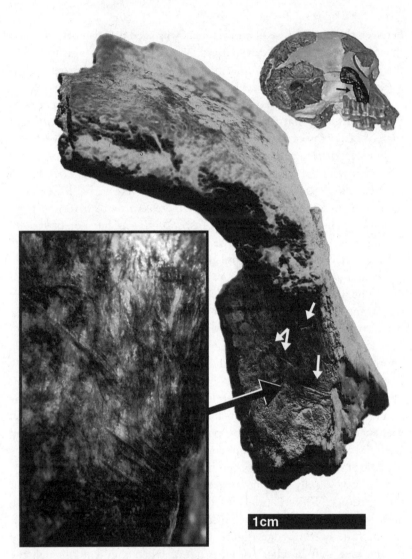

**FIGURE 11.** The butchered remains of StW 53, an ape-man fossil from Sterkfontein Cave, South Africa. The rendering in the upper right corner shows the whole cranium (dark areas represent the actual bone fragments that were recovered, white areas are reconstructed sections linking those fragments), with the cut-marked cheek bone indicated by a black arrow. Moving down and toward the left, the cheek bone is shown enlarged, with arrows indicating discrete clusters of cut marks. The bottom left shows some of the cut marks at even higher magnification. (Photograph courtesy of Tim White and the Human Evolution Research Center, University of California, Berkeley; used by permission of John Wiley & Sons, Inc.)

I pause here to note that Ron Clarke has argued vigorously with me that our interpretation of the damage on StW 53 is, no pun intended, off the mark. Clarke points out the relevant fact that the deposit from which StW 53 was excavated contains naturally occurring pieces of quartz and chert, many with sharp edges. He ventures further that

> sharp quartz and chert can be forced against bone surfaces either during [the formation of bone assemblages on sloping debris cones, as sediment is accumulating in a vertical opening to a cave; see chapter 2] or during collapse episodes. . . . Thus one could question whether cut marks made by a hand-held [stone tool] would differ from cut marks made by natural stone forced against the bone in a debris slope. In fact, the cut marks on StW 53 were, I believe produced naturally by a small chert block in the area of the zygomatic arch and which moved against the bone under pressure in the [debris] slope. A block was there when the cranium was discovered and can be seen in a cast made at the time.

If Clarke is correct, this means that the striations White, Toth, and I observed on StW 53 are not real cut marks, but instead pseudo–cut marks—the result of a (nonbiological) geological process rather than of hominin behavior. The difference holds great significance for reconstructions of early hominin activities at Sterkfontein.

Ron is a close friend and I also have profound respect for him as a scientist. So, for at least two reasons, I considered his misgivings very seriously. More than that (as discussed earlier), I am part of the research group that expressed a similar skepticism about the purported 3.4 million-year-old cut marks from the Dikika site in Ethiopia. I try always to keep an open mind, ever cognizant that the events we paleoanthropologists are reconstructing happened a very long time ago, and in the absence of the direct observations of any of us. Misinterpretations of the evidence are bound to happen—and, although those misinterpretations are often easy to detect when they emanate from a colleague, they are sometimes extremely difficult to see and acknowledge when they are one's own. It is, however, the willingness to accept the falsification of one's preferred hypothesis that marks a scientist who is more concerned about capturing the reality of the past (or, at least, coming as close to that reality as is currently humanly possible) rather than with advancement of a personal agenda. So, after absorbing Ron's doubt, I often went back to StW 53 in my mind and my notes, and once in the laboratory. Upon careful reexamination of all the constituent fragments that make up the skull, I still do not discern trampling damage on any of them—including even those fragments that because of their shapes would have been far more likely to have incurred random abrasive damage than was the cut-marked piece of cheek bone. In addition, my analysis of the complete fossil assemblage, of which StW 53 is simply a small part, demonstrated the acute rarity of trampling damage in the collection as a whole.

But, in fairness to Clarke, he is not arguing for trampling marks on StW 53, but more specifically that because "the cranium was slightly crushed, [it] suggests the strong likelihood that a sharp chert block was forced against the bone in question during the [geological] events that crushed the cranium through the pressure of sediment load." Ironically, years ago, I argued in my doctoral thesis that a similar geological process created several provocative striations on a fragment of *Australopithecus* lower jawbone from a portion of Sterkfontein that is even older than the 2 million-year-old StW 53 level. In fact, Clarke cites the damage on this jawbone to support his counterargument to White's, Toth's, and my claims about the marks on the StW 53 skull. The marks on the older *Australopithecus* jawbone are, however, of a different character than those on StW 53. Whereas the StW 53 striations conform to cut marks in their location on the skull (that is, they occur where one would expect a butcher to accidentally leave them when cutting meat from the face and disarticulating the lower jaw from the cranium), and in their macro- and microscopic morphologies, those on the older ape-man jawbone bear very little resemblance to damage inflicted by stone tools. To this point, in 1999 I wrote:

> The context of the [lower jawbone] has led to the interpretation that the [marks were] caused by dynamic impact with [or crushing by] the jagged surface of a rock in antiquity. First, the other side of the [lower jawbone, without marks] has been crushed . . . as has the right [upper jawbone]. Second, there is no reason to believe that the marks were inflicted intentionally by a stone tool, as they are much shallower and less steeply sided than typical cut marks made by an artifact. Third, the damage is unlike striations caused by modern excavation and preparation tools.

As in the case of the suggestive Dikika bone surface marks, I was also unable to reject the null hypothesis of a natural (that is, non-hominin, non-butchery) origin for the marks on the *Australopithecus* jawbone.

My contrasting observations of damage on StW 53 and on the older *Australopithecus* jawbone do not mean that Clarke's interpretation of the marks on StW 53 is wrong, but only that, in my opinion (always subject to change, provided I am presented with additional evidence to the contrary), our conclusion—that StW 53 is cut-marked by stone tools—is currently the "best fit" hypothesis. Thus, I still agree with our original report on the damage to StW 53's facial skeleton, including the carefully worded conclusion: "It is not possible to infer the reason(s) for the cut marks observed on the StW 53 hominin specimen. Reasonable hypotheses include cannibalism, curation, mutilation and/or funerary practices." The most convincing cases in which prehistoric cannibalism has been deduced are based on demonstrations that hominin remains were treated by other hominins in the same manner as were the co-occurring bones of non-hominin animals. Critical thinkers should have little problem with the supposition that hominin bones scarred by defleshing cut marks and percussion marks, and then dumped unceremoniously in the site, represent refuse from hominin meals—just as is inferred for animals also deposited at the site in the same condition. Proponents of cannibalism hypotheses argue that rejecting cannibalism in these cases, in which there is identical rough treatment of hominin and animal bones, is to deny, de facto, that we can make any inferences at all about early hominin diet from studies of butchered bones.

I agree. Unfortunately, though, the bone assemblage from the StW 53 deposit is very small, so building the contextual case for cannibalism is not possible. In fact, the *only* cut-marked specimen from the StW 53 deposit is the hominin cheek bone. That said, if I had to choose between the "reasonable hypotheses" offered above for the StW 53 cut marks, cannibalism seems the *most* reasonable. My rationale for that choice is based on the deep antiquity from which StW 53 derives— again, it is around 2 million years old—and because the frequency and intensity of cut marking on it is moderate, indicating that the damage was probably inflicted unintentionally during the course of normal, subsistence-motivated butchery. Also, there are no other convincing indications of symbolic behavior in the paleoanthropological record at this early time; in that context, ritualistic modification of the StW 53 cranium seems very unlikely.

It is not until about 600,000 years ago, with a butchered hominin cranium from the Bodo site, in Ethiopia, that cut marking beyond what I would expect from simple diet-related butchery is evident. Tim White documented seventeen distinct clusters of cut marks across the entirety of the Bodo cranium, underlining what a crazed mess its butcher created those hundreds of thousands of years ago; the various positions of the cut marks indicate scalping, gouging out of the eyeballs, and removal of the face (in testament to its shock value, a newspaper report of the savagery evinced by Bodo serves as an epigraph to the great American novel *Blood Meridian*, Cormac McCarthy's classic discourse on violence and horror). And, it is only much more recently, about 160,000 years ago, that the archaeological record shows indications of the probable curation by hominins of the defleshed skulls of other hominins. Stone tool cut marks scar the crania of an early modern human adult and child recovered from the Herto site, in the Middle Awash, Ethiopia. The child's cranium is also polished, with a highly visible postmortem sheen built up on its sides. Some modern New Guinean tribes venerate their dead ancestors by flaying, decorating, and keeping their skulls. With all the handling and shifting around that they endure, these ritually prepared skulls eventually take on gloss distinctly similar to that on the Herto child's cranium.

It seems likely that Raymond Dart would have reveled in the ghoulish disclosures from the Bodo and Herto fossil fields—and would have also delighted in our interpretation of the marks on StW 53, confirming his view that hominin cannibalism, consistently evident in more recent phases of human evolution, had its origin deep in our prehistory. But was it *technically* cannibalism at Sterkfontein? The possibility exists that another kind of hominin, and not StW 53's species of *Australopithecus*, was the butcher. We know, for instance, that early *Homo* had evolved by the time of StW 53. Could a technologically advanced *Homo* cousin of StW 53 have foreshadowed the sad and recurring story of modern human colonists decimating materially disadvantaged aborigines when the two came into contact? We'll never know, but insights about Dart's own Taung Child urge that we consider a much fuller range of nasty and brutish insults that *Australopithecus* confronted day to day in its hardscrabble world.

## BLITZ

Not all danger emanated from the intellectual superiors of *Australopithecus*. Fossils don't easily reveal the sundry snake bites, heat exhaustion, and viral infections undoubtedly suffered by the ape-men. Bones do, however, show that our ancestors were the frequent prey of large carnivores. Bob Brain's observation of leopard canine punctures in the skull of a child robust australopithecine from Swartkrans Cave (see chapter 4) comes readily to mind. And, since that early demonstration, other research has revealed carnivore tooth marks on *Australopithecus afarensis, anamensis,* and *africanus* fossils, as well as on the more ancient hominin remains of *Orrorin* and *Ardipithecus*. Even large monkeys, with canine teeth much bigger than those of early hominins, would have been formidable competitors and potential killers of ape-men, when a shared fondness for the same fruits and other foods probably drew the two groups into regular contact and potential conflict.

The fragility of ape-man existence is, however, personified most dramatically by the Taung Child. The Taung site is unique in South African paleoanthropology because its child is the only hominin fossil recovered there; other caves, like Sterkfontein and Swartkrans, have yielded the remains of *hundreds* of individual ape-men. Raymond Dart and Robert Broom both commented long ago that Taung is also exceptional in its abundance of small animal fossils, like those of lizards, monkeys, small antelopes, and rodents; bird eggshells, tortoise shell fragments, and parts of crabs are also unusually common at Taung. But, it was not until much later, in 1995, that paleoanthropologists Lee Berger and Ron Clarke offered a compelling explanation for the singularity of the Taung fossil assemblage. They noted that eagles tend to prey on the very types of animals found at Taung, including significantly, small primates. When an eagle eats, the bones of its prey often fall out of its nest and accumulate on the ground below. If an eagle's nest was located above the cave opening at Taung, this taphonomic process could certainly explain the accumulation of the small animal parts in the site. But, Berger and Clarke went further, estimating that when alive the Taung Child weighed about 20 pounds, well within the body size range of prey targeted by eagles. The mounting implication is chilling: the Taung Child, a three- or four-year-old toddler, separated—perhaps just momentarily—from his mother was plucked away into thin air by an unseen, overhead assailant.

Incredulity is the next reaction. One can easily envisage a rabbit or even a monkey being picked off by some wretched bird—but could *Australopithecus*, our own ancestor, have been just as feckless when confronted by these menacing airborne overtures? The historical record contains anecdotal reports of large raptors capturing and flying away with human children, and Berger and Clarke refer to more recent examples of crowned eagles, one of the largest of the extant African species, harrying people: "In Zambia crowned eagles have even been recorded attacking, and nearly killing, a seven year old human child of approximately [45 pounds] . . . , and there are numerous reported cases of adult humans being attacked while near nests." And, in Zimbabwe a child's skull was recovered from the nest of a crowned eagle, a sad forensic clue that eagle interaction with people, even today, might go far beyond simple harassment.

Finally, damage on the Taung Child's skull (scratches and punctures from an eagle's talons and beak) corresponds to that on monkey skulls recovered from modern eagle nests (figure 12). Validation of the killer ape is certainly not found on the bones of Dart's unfortunate child, who, in the innocence of his frailty 3 million years ago, seems to ridicule his discoverer's ferocious intellectual spirit. That judgment, of course, misses the point. Like every organism, past and present, *Australopithecus* was an indefinable amalgam of capability and vulnerability. Grasping for its disposition remains one of paleoanthropology's greatest challenges.

That is why Owen Lovejoy's model of early hominin sociality—love it or hate it—stands as an important theoretical crux from which the twisted conflation of human aggression and human hunting might, somehow, someday, be disentangled and each behavior assessed on its own. If Lovejoy's hypothesis is correct, then general emotional

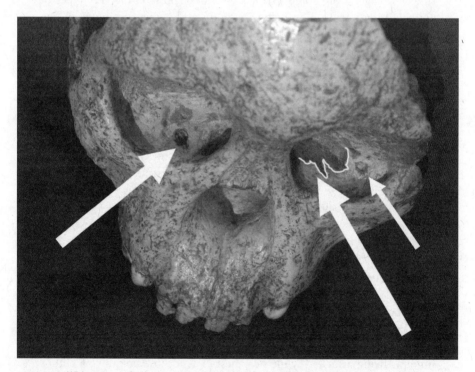

**FIGURE 12.** White arrows indicate punctures and broken bone in the ocular sockets of the *Australopithecus africanus* Taung Child cranium, where an eagle, in the course of killing and eating the child, raked and gouged out his eyes. (Photograph courtesy of Lee Berger and the University of the Witwatersrand; used by permission of John Wiley & Sons, Inc.)

control—a prerequisite for the success of modern human hunting—was probably established very early in the prehistory of our lineage. Indeed, the small, feminized canine teeth of male *Ardipithecus* stand as proxy witness to the likelihood of reduced intragroup aggression and increased cooperation among males in this purported earliest hominin. Thus, the human *capacity* to disassociate aggression from hunting was almost surely well established by 200,000 years later, at the beginning of the age of the ape-men, the apparent descendants of *Ardipithecus*. Because the paleoanthropological record is incomplete, the technological *capability* necessary for our uniquely human style of hunting will never be documented definitively. But, clues—such as woodworking (spear making?) traces on stone tools, hominin brain expansion and gut reduction, abundant butchery marks on prime, meat-bearing bones of large ungulates, and the colonization of temperate, seasonally plant food–poor Eurasia—indicate that this capability was in place possibly as early as 2.5 million years ago with *Australopithecus garhi*, and surely by 1.8 million years ago with *Homo erectus*, when the fossil and archaeological records show a profusion of these indicators of human-like hunting.

No animal that ever existed holds more destructive antagonistic potential than does *Homo sapiens*. Dart recognized this more than sixty years ago and was so impressed by the awesome realization of it that he coupled aggression and big-game hunting—one of the defining capabilities of modern humanness—in order to model our emergence from a troglodytian root. Indeed, it seems that Dart was entirely unable to appreciate that modern humanity's nigh–death wish coexists but is unrelated to the contrasted placidity of a traditional hunter, whose success is dependent on calm restraint.

It now seems apparent that the way forward in paleoanthropology—understanding ourselves and our direct ancestors as the really unusual primates that we are and were—requires that we disaggregate notions of aggression and hunting. When Dart postulated a "predatory transition from ape to man" in 1953 he was ignorant of chimpanzee hunting, details of which emerged only nine years later from Jane Goodall's work at Gombe. But, in a sense, Dart's idea *was* prescient. Setting aside the unusual case of bushbaby hunting by chimpanzees at Fongoli, predation by apes and by humans is a study in contrast. And, because subsistence is such a fundamental aspect of an organism's biology, that contrast takes on great significance in distinguishing ape from human. As Katharine Milton and many others posit, increasing its intensity of meat eating was pivotal to the full humanization of evolving *Homo*. Human hunting style—taking large game by the unique combination of distance-providing technology (Dart was prescient here too, arguing that extrasomatic weaponry was a hallmark of humanness), intellect, and emotional control—permitted that regularization of meat eating.

Our prehistoric ancestors held no inexorable sovereignty on the savanna, but neither did they grovel before some imagined "betters." Almost always outmatched physically by both their predators and their larger prey, early hominins played the cards they did hold with great acumen. Resolve, brain power, and composure ultimately ensured their survival and our eventual existence. Just as with modern people, a dark impulse must have simmered shallowly within the ape-men and *Homo erectus*, but also, just as with modern people, its expression was only one of the myriad of human expressions. They and we are linked—not simply as ancestors and descendents, but also in our all-encompassing capacity for both the lowly and the majestic.

# Coda

Sir, I guess there's just a meanness in this world.

—Bruce Springsteen, "Nebraska"

Preachers and politicians pontificate about the root of human violence. Philosophers and psychologists intellectualize its provenance. Collectively, their views span erudite heights to ridiculous lows—sometimes inaccessible, sometimes commonsensical, and sometimes laughable. But, none of those learned (or conniving) voices comes close to writer James Ellroy in explicating so brusquely and so beautifully the wellspring of small-scale violence (forget terrorism, war, acts of self-preservation, serial killers, and other deranged murderers here). In his more specific meditation on (violent) crime, Ellroy writes:

> You had to control. You had to assert. It got crazy and forced you to capitulate and surrender. Cheap pleasure was a damnable temptation. Booze and dope and random sex gave you back a cheap version of the power you set out to relinquish. They destroyed your will to live a decent life. They sparked crime. They destroyed social contracts. The time lost/time-regained dynamic taught me that. Pundits blamed crime on poverty and racism. They were right. I saw crime as a concurrent moral plague with entirely emphatic origins. Crime was male energy displaced. Crime was mass yearning for ecstatic surrender. Crime was romantic yearning gone bad. Crime was the sloth and disorder of individual default on an epidemic scale.

This is a mostly Lorenzean, instinct-centered view of the inner killer ape finally snapping its human manacles: the cap blown off an overheated radiator. It is a view that posits that most small-scale violence is simply impulse detonating in men who are affected by the perfect internal alchemy of testosterone and inhibition-reducing substances. The specific effects of interpersonal violence are often devastating to victim and perpetrator alike, but the underlying cause of that kind of violence and its action are ultimately pedestrian and entirely predictable. Sin pulsates just below the surface in each of us, continually testing against the inside of our ribcages. Conrad's heart of darkness. Whether that innate human aggressiveness is, in the Lorenzean understanding, a vertebrate instinct honed by natural selection over the eons, or, instead, in Freud's view, each person's internalized death drive, matters little for understanding the psychology of war and the other machinated horrors spawned by humanity upon humanity. Lorenz and Freud agreed that the societally imposed repression of innate human hostility causes mental distress, "discontent" in the individual. But, they diverged in Lorenz's view that this means society itself is sick versus Freud's renowned proposition that society, civilization, is the retaining wall pushing back the sickness that would otherwise burst from each of us. In either case, humans, because they are innately aggressive, are condemned to wage war. To these thinkers, it is altogether appropriate to extrapolate upward from the individual.

This bottom-up way of understanding war predates Lorenz and Freud by centuries and more. Its exact parentage is lost in the ragged palimpsests of time before history, but a well-known example of the sentiment from the Age of Reason is found in the writings of Joseph-Marie, Comte de Maistre (1753–1821). De Maistre, an apologist for absolutism and stooge of the Counter-Enlightenment, is no hero of mine or of any freethinker. Nonetheless, the following quote proves him not only prophetic in appreciating the centrality of heterotrophy to the evolution of complex life (see chapter 1) but also as another link in the long chain of important thinkers confounding (albeit, in de Maistre's case, eloquently) subsistence predation with interpersonal violence, war, and human dominionism:

> In the immense sphere of living things, the obvious rule is violence, a kind of inevitable frenzy which arms all things in mutua funera. Once you leave the world of insensible substances, you find the decree of violent death written on the very frontiers of life. Even in the vegetable kingdom, this law can be perceived: from the huge catalpa to the smallest grasses, how many plants

die and how many are killed! But once you enter the animal kingdom, the law suddenly becomes frighteningly obvious. A power at once hidden and palpable appears constantly occupied in bringing to light the principle of life by violent means. In each great division of the animal world, it has chosen a certain number of animals charged with devouring the others; so there are insects of prey, reptiles of prey, birds of prey, fish of prey, and quadrupeds of prey. There is not an instant of time when some living creature is not devoured by another.

Above all these numerous animal species is placed man, whose destructive hand spares no living thing; he kills to eat, he kills for clothing, he kills for adornment, he kills to attack, he kills to defend himself, he kills for instruction, he kills for amusement, he kills for killing's sake: a proud and terrible king, he needs everything, and nothing can withstand him. He knows how many barrels of oil he can get from the shark or a whale; in his museums, he mounts with his sharp pins elegant butterflies he has caught in flight on the top of Mount Blanc or Chimborazo; he stuffs the crocodile and embalms the hummingbird; on his command, the rattlesnake dies in preserving fluids to keep it intact for a long line of observers. The horse carrying its master to the tiger hunt struts about covered by the skins of this same animal. At one and the same time, man takes from the lamb its entrails for harp strings, from the whale its bones to stiffen the corsets of the young girl, from the wolf its most murderous tooth to polish frivolous manufactures, from the elephant its tusks to make a child's toy: his dining table is covered with corpses. The philosopher can even discern how this permanent carnage is provided for and ordained in the whole scheme of things. But without doubt this law will not stop at man. Yet what being is to destroy him who destroys all else? Man! It is man himself who is charged with butchering man.

It is hard to deny that an individual willingness to wage war must, at a deep level, exploit our innate potential for aggression. The journalist and social philosopher Chris Hedges maintains that war, for the individual, is transcendent. In Hedges's view, war not only provides a sense of purpose in a life otherwise without purpose—but, it also looses our deepest primate lusts. It is Freud's death instinct liberated under a guise of legitimacy, acting at the behest of our elected rulers and cloaked in false patriotism and sham morality. In war, cessation of control over our basest internal impulses is allowed, authorized— indeed, compelled—by the very state that we serve in the execution of its corporatist ambitions and hostilities.

Small-scale violence also concedes control—but without legal consent or legal encouragement. There is no facade of purposefulness and no premeditation in small-scale violence. Small-scale violence is erratic— and, because so, it is an inept way for a human to hunt large, wary, and sometimes dangerous prey. Good human hunters don't channel their inner killer apes; they suppress them—and to evolutionarily significant effect. Most often, nondefensive, small-scale violence is exhausting and entirely counterproductive. Men in our society who don't learn this lesson early on become jailhouse habitués or artful dodgers. As for foragers: show me a Hadza hunter who crashes through the underbrush like a wrecking ball, wild-eyed-ready to slake a bloodlust, and I'll show you a guy without a meal, without marital prospects, and without much hope of taking care of himself for very long.

Of course, it's hyperbole to contend, by extension, that small-scale violence and tool use are completely antithetical. Even compulsive murderers can kill with a bludgeon or a shank. However, those tools, affecting those passionate acts, are usually expedient and makeshift—they do little to remove the killer from dangerously close proximity to his victim. Hunting tools are entirely different, the products of intent and ingenuity.

Coming to terms with our inherent potential for violence is critical for a host of sociological reasons and moral concerns, but the lessons revealed in that pursuit will reflect little light on the evolution of hominin hunting and its pivotal role in our becoming human. Even so, it seems quite unlikely that human hunting, interpersonal aggression, and war will ever be fully disentangled intellectually. But, perhaps we are approaching a new era in the science of human evolution in which at least the flippant conflation of these phenomena will cease. The influential work of the economist Samuel Bowles is a recent case illustrating a new, more nuanced view of the relationship between hunting and war. In an effort to understand the evolution of altruism among presumably selfish individuals (shaped by and subject to evolutionary forces), Bowles, and his frequent collaborator Herbert Gintis, simulated the socioeconomic dynamics of ancient human groups. Based on their results, they proposed "that the crucial step towards [modern] human social systems was the evolution of a cooperative unit that was big enough *to ensure against the risks involved in hunting large game*, comprising around thirty-two adults plus juveniles and the elderly" (my emphasis). Thus, according to Bowles and Gintis, human altruism, prosocial behavior, cooperation—call it what you will—was a product of the Pleistocene and of predatory concerns.

Bowles did recognize, however, that there would have been so little genetic variation between individuals in nascent social groups of this small size that

> even if violent conflict between bands was common, group selection could
> not favour costly altruistic acts [that is, engaging willingly in warfare

conducted at the level of potential self-sacrifice of life by fully aware, "altruistic warriors"]. Selection among groups for cooperation [with altruistic warriorism—because of its high potential to be fatal to the warrior—among the most extreme of all cooperative traits] gains traction only where it exceeds that for selfishness within a group. Social institutions can reduce the advantages of selfish behaviour. For example, modern hunter-gatherers typically share their resources. More successful providers are "taxed" in food to support the collective. This sharing limits variation in reproductive success within groups, imposing selection on groups for individually costly acts of cooperation.

From this perspective, Bowles and Gintis can explain the existence of many nearly universal human mores and norms (like sharing and xenophobia), as well as the social sanctions (like ostracism, expulsion, and corporal punishment) that institutionalize and maintain those mores and norms. But, Bowles goes further still. Reviewing ethnographic and archaeological evidence of intergroup conflict, he contends that warring was a frequent and sustained part of human life in many places in the recent past. Under those circumstances, and with hypothesized increases in primitive population sizes, Bowles then argues that the quantitative relationship of within-group genetic variation and between-group variation was such that costly (indeed, even potentially lethal) altruistic behavior—including altruistic warriorism—might have been selected.

There are some important weaknesses in Bowles's model: the archaeological evidence of early warfare is scant and ambiguous; Bowles is dealing with, relatively speaking, quite recent stages of human prehistory (at the most, only hundreds of thousands, and not millions, of years ago); some researchers question the underlying assumption, on which his model is dependent, of elevated genetic differences in neighboring human groups. These, among others, are significant caveats. Still, they do not diminish the audacity of Bowles's endeavor to provide a unified model of human cooperation and to explain fundamental human behaviors that are, at the same time, seemingly incongruent with evolutionary expectations of ourselves as selfishly acting individuals who should be first and foremost concerned with assuring our own survivorship and genetic replication.

Regardless, the point in outlining Bowles's ideas here is to emphasize that he, like so many before him, concludes an underlying causal relationship between hunting and violence. Follow Bowles's thread back in time, and war, in all its manifold expressions and complexities, is ultimately germinated from the seed of cooperative hunting. But Bowles's view of the relationship between war and hunting is nuanced, distantly stepwise, and wholly contingent. Unlike Raymond Dart and Richard Wrangham (and Charles Morris, Harry Campbell, and Carveth Read, before Dart and Wrangham), Bowles does *not* equate hunting and war as twin blights oozing from some immutable primate disposition for mayhem. Instead, Bowles seems to recognize the lability of human nature—human nature that is as likely to give rise to love as it is to brutality.

Love and brutality: emotional states both fraught with thrill.

Likewise, hunting holds much thrill. But, for humans, it is thrill that must be checked. The good hunter's self-denial, suppressing his urge toward giddiness, yields rewards that are expressed proximately in post-hunt states of exuberance, satiation, and adulation. But, the ultimate evolutionary payoff of emotionally checked hunting is the increases that the good hunter gains in his inclusive fitness: because of his composure, he lives to eat and to pursue successful reproductive opportunities with mates who will be invested, hard-working mothers. Further, his success provides for these women and his offspring with them, children who are the man's direct evolutionary legacy.

If at all, hunting is only the haziest expression of human nature. Our inherent potential for small-scale interpersonal violence, which, in the absence of mental imbalance, is usually harnessed for self-defense and to assert dominance and to avert being dominated, provides little comparison beyond the mechanical for understanding human predation—and even less for understanding full-scale warring. Without advanced, modern technology, there are only so many ways for one human being to destroy another animal, be that other animal prey, victim, or enemy. But, hunting is for food, and primeval war for other evolutionary benefits. And modern, state-level warfare is, for the individual warrior, completely divorced from both. As Hedges so clearly recognizes, modern rulers leverage the individual's deep-seated biological lusts and his desire for belonging and purpose in order to create a warrior who then, through war, advances the rulers' political aspirations and corporate interests. Although underlaid by human nature, this is stuff far removed from our ancient roots and evolutionary pressures. Anthropologists, as well as all people, are wise to heed the disparity. Capturing the realism of the past depends, in large part, on the distinction—as does rectifying the transgressions of the present and staunching those of the future. Darwin and Dart both appreciated that we will never rid ourselves of the darker angels of our nature, but the place to grapple with them is in this world and not in that of the past.

# Notes

I endeavored to keep my notes concise, citing only the major sources that I consulted (complete bibliographic information for those sources is found in the References). The notes do, however, contain much information that is of socio-historical and scientific importance; the inclusion of this material in the main text would have encumbered the chapter narratives in a series of intellectual excursions. If you are reading this introduction to the notes, you are probably primed and ready for just such excursions, and thus, I wish you happy reading, with the hope, and in anticipation, that you will find much stimulation in the following pages.

## INTRODUCTION

This is not a book about human nature per se. But it is a book that accepts the notion that there is, indeed, such a thing as human nature and that fully understanding the course of our evolution requires an appreciation of its influence on that trajectory. Of course, this approach is now quite common in the behavioral sciences—becoming especially well fixed in the discombobulating wake of E. O. Wilson's sociobiological onslaught that took hold in the mid-1970s. For all its many bald scientific missteps since Wilson's revolution, sociobiology and its academic congener, evolutionary psychology, can lay claim to a deep and renowned intellectual parentage. The lumping together of Hobbes, Locke, and Rousseau as "blank slaters" does not categorize but only caricatures these giants of the Enlightenment and Romantic periods; in actuality, each viewed various forms of innate human selfishness as underpinning the accreted nature of humans—be that increscent nature, under their various, fully articulated constructs, "good" or "bad." But, more important than merely recognizing the blatant simplification of intricate ideas about human nature, modern psychology and neurobiology have, in the ensuing centuries, pressed beyond these essential philosophers and marshaled an impressive body of empirical data falsifying hypotheses of the tabula rasa.

Wilson's (1975) classic treatise on sociobiology is *Sociobiology: The New Synthesis;* for antecedent and decisive works in the field, see also: Hamilton (1964); Trivers (1972, 1985); Alexander (1974, 1987); Wilson (1978). Seminal work in the allied field of evolutionary psychology is found in Symons (1979) and Barkow et al. (1992). Steven Pinker's (2002) *Blank Slate: The Modern Denial of Human Nature* provides a good review of the biological basis of human nature; Pinker also explores the societal and political implications of acknowledging human nature. Michael Shermer's (2004) *The Science of Good and Evil* summarizes an impressive range of behavioral, cognitive, and neurological data supporting the assertion of the malleability of human nature; more dubious are the proposed relationships of these data to human morality. One of the best recent books on the relationship between science and morality is Sam Harris's (2010) *The Moral Landscape: How Science Can Determine Human Values.* The quote on page 1 is from Boehm (1999). The quote on page 3 is from *Under Western Eyes* (Conrad, 1911).

**CHAPTER 1: A MAN AMONG APES**

Bob Brain's research on the world's earliest animals continues (Brain et al., 2001, 2003, 2012). Histories of the science of paleoanthropology can be found in Lewin (1997), Spencer (1997), and Delisle (2006). Craig Stanford's (1999) popular account, *The Hunting Apes: Meat Eating and the Origins of Human Behavior,* followed *Chimpanzee and Red Colobus: The Ecology of Predator and Prey,* his technical text on chimpanzee hunting from 1998. Wrangham and Peterson (1996) is an engaging read about the "demonic male hypothesis," and Hart and Sussman (2005) explain their ideas about early human susceptibility to predation. The quote on pages 7–8 is from White (2006).

*Does the Boy Make the (Hu)man?*

See Burnett (1773–92) and Darwin (1871) for their early views of the close relatedness of humans and apes. The story of Eugène Dubois and his discovery of the first known *Homo (Pithecanthropus) erectus* fossils is told by Theunissen (1989) and Shipman (2001). Susan Antón (2003) gives a detailed overview of the inferred natural history of the species; see also Dunsworth and Walker (2002). Brain size estimates for *Homo erectus* are from Holloway et al. (2004). A thorough and very technical volume, edited by Walker and Leakey (1993), provides details of Nariokotome Boy's discovery, his preserved anatomy, and initial inferences about his growth and development and behavioral capabilities; the skeleton's announcement and first description are by Brown et al. (1985). An engaging popular account of the Nariokotome Boy skeleton and its scientific significance is also available (Walker and Shipman, 1996). Walker's quoted opinion about the humanness of Nariokotome Boy (page 15) comes from this latter book.

The contention that Nariokotome Boy's long, linear body was physiologically adapted to his tropical African habitat is supported by a well-known biological principle called Allen's Rule. Allen's Rule is based on empirical observations that long limbs increase an endotherm's (that is, a warm-blooded animal's) body surface area without increasing its body volume significantly. Mammals with relatively large body surfaces and relatively small body volumes dissipate heat more rapidly and efficiently than do stocky, short-limbed mammals, who, in contrast, have relatively large body volumes but relatively small surface areas. Recent research by Graves et al. (2010; see also Ohman et al., 2002) contends that Nariokotome Boy "would have experienced a growth spurt that had a lower peak velocity and shorter duration than the adolescent growth spurt in modern humans. As a result, it is likely that [Nariokotome Boy] would have only attained an adult stature of 163 cm (~5'4"), not 185 cm (~6'1") as previously reported." In addition, the anatomy of a 1.4 million–900,000-year-old hominin pelvis from Busidima, in Ethiopia, would also seem to contradict the idea that the body of *Homo erectus* was adapted to semi-arid tropical environments. Simpson et al. (2008) contend the Busidima pelvis is from a female *Homo erectus* who was short and broad hipped. Based on this morphology, Simpson and his colleagues argue further that, as a long-lived and geographically widely distributed species, *Homo erectus* did not yet "exhibit the same ecogeographic patterns of body form as seen in modern humans"—that is, the occurrence of at least one diminutive, thick-set *Homo erectus* individual from Busidima violates predictions of Allen's Rule as now expressed in modern humans, with, for example, rangy Dinka found in the Sundan and squat Inuit distributed across the northern latitudes of the New World. Complicating the story, Ruff (2010) contends that the Busidima pelvis might not be that of *Homo erectus* but might instead belong to the robust ape-man species, *Australopithecus boisei*, an East African contemporary of *Homo erectus*.

Bogin (1999) and Bogin and Smith (2000) provide good overviews of the pattern of human growth and development and consider how and when the human life cycle evolved. More detail on the timing of modern human tooth development and eruption can be found in Anderson et al. (1976) and Liversidge (2008). General discussion of limb bone growth and development is found in White, Black, and Folken's (2009) essential anatomical text, *Human Osteology*, and in more detail in Ogden (1990) and in Scheuer and Black (2004). Holly Smith (1993) provided the first comprehensive analysis of the physiological age of Nariokotome Boy. It was Dean and Smith (2009) who came to the remarkable conclusion that the Nariokotome Boy died at a much younger age than was previously estimated.

Note that the teeth and skeletons of humans develop as separate processes (see, e.g., Lewis, 1991; Smith, 2004), so a "mismatch" between the Nariokotome Boy's estimated ages of death based on the developmental state of his teeth and that of his skeleton was not unexpected. However, the large discrepancy of nearly three years between the teeth- and skeleton-based estimates *was* surprising; the disparity is "outside 99 percent limits for normal [modern human] children and well into the distribution for growth disorders" (Dean and Smith, 2009; see this reference also for a good summary and discussion of the relevant modern comparative data).

See Hillson (1996) for a good general overview of dental anthropology and specific discussion of perikymata and human tooth development. Christopher Dean's (for example, Dean, 2000, 2006; Dean et al., 1993, 2001; Beynon and Dean, 1988) work on these issues in paleoanthropology is the gold standard in the field. Additional relevant and interesting contributions from other workers on early hominin taxa include Smith (1986, 1992), Beynon and Wood (1987), Kuykendall (2003), Moggi-Cecchi et al. (1998), and Lacruz et al. (2006, 2008).

MacLarnon and Hewitt (1999, 2004) discuss the importance of breathing control for advanced human language capability. MacLarnon's (1993) interpretation of the significance of the narrow thoracic vertebral canal of Nariokotome Boy has been challenged by Latimer and Ohman (2001), and this challenge subsequently countered by MacLarnon and Hewitt (2004).

### The Gut of the Matter

The earliest known modern human (*Homo sapiens*) fossils are from the sites of Kibish (~195,000 years old) (McDougall, Brown, and Fleagle, 2005, 2008; Brown, McDougall, and Fleagle, 2012) and Herto (160,000–154,000 years old) (Clark et al., 2003), both in Ethiopia; the oldest fossils of the genus *Homo* (2.3 million years old) are from Hadar, Ethiopia (Kimbel, Johanson, and Rak, 1997); the oldest fossils of *Australopithecus* are assigned to *Australopithecus anamenis*, from sites in Kenya and Ethiopia (Leakey et al., 1995, 1998; White et al., 2006). The Nariokotome Boy skeleton, at about 1.8 million years old, is among the oldest known *Homo erectus* fossils, but a fragment of hominin braincase (like Nariokotome Boy, from the vicinity of Lake Turkana in Kenya) might also be *Homo erectus* and is ~100,000 years older than is the boy's skeleton (Gathogo and Brown, 2006). Early *Homo erectus* fossils, nearly 1.8 million years old, also known outside of Africa, at the Georgian site of Dmanisi (see references at the end of this section), and archaeological evidence of *Homo erectus* at Dmanisi predates its fossil evidence by almost one million years (Ferring et al., 2011). See text and notes to chapter 2 for discussion of the earliest putative hominins, *Ardipithecus*, *Orrorin* and *Sahelanthropus*.

See Katharine Milton (1999 [source of quote on page 16], 2003) for good synopses of her view of the importance of meat eating in human evolution. Milton's idea that meat both satisfies many dietary requirements *and* allows for increased reliance on plants with relatively low nutritional but high energetic content—required for the large bodies of early *Homo*—are elaborated upon by Aiello and Wells (2002; see also Aiello, 2007; Aiello et al., 2001) as part of Aiello's overall defense of Aiello and Wheeler's (1995) "expensive tissue hypothesis." The old prevailing view, that *Australopithecus* had an apelike thorax, was based largely on the work of Schmid (1983). Reconstruction of the thorax of the Woranso-Mille *Australopithecus* skeleton is presented in Haile-Selassie et al. (2010). Bruce Latimer, Scott Simpson, and Yohannes Haile-Selassie (2011) provided more details on the Woranso-Mille ribcage in a presentation at the 2011 Annual Meeting of the Paleoanthropology Society, entitled *Thoracic Shape in* Australopithecus. The presenters made the case that the restricted upper ribcage of African apes is due to knucklewalking, their specialized form of terrestrial locomotion. If so, then the expanded upper ribcage of early *Australopithecus* indicates that knucklewalking was never part of the locomotor repertoire of hominins. Ward et al. (2012) describe the new thoracic vertebrae (and other fossils) of *Australopithecus afarensis* from Hadar. The quote on page 18 is from Navarrete et al. (2011).

The remarkable finds of early *Homo* at Dmanisi, in the Republic of Georgia, are described in Gabunia and Vekua (1995), Gabunia et al. (2000), Vekua et al. (2002), Lordkipanidze et al. (2005, 2006, 2007), and Rightmire, Lordkipanidze, and Vekua (2006). Lordkipanidze et al. (2005) document the evidence of hominin butchery—in the forms of sharp-edged stone cutting tools and butchered bones—at Dmanisi, and Shipman and Walker (1989) outline the physiological reasons why a considerable degree of hominin carnivory was almost certainly necessary in order for hominins to successfully colonize high, temperate latitudes.

### CHAPTER 2: PREHISTORIC BLOODSPORT

There are a few biographies of Raymond Dart. None of them is of very high quality, but, just the same they are, in chronological order: Wheelhouse (1983); Tobias (1984); Wheelhouse and Smithford (2001). Unsurprisingly, the most entertaining sketch of this important figure's life is his autobiography (Dart and Craig, 1959), heavy on anecdote and, at turns, charming and excruciating in the floridity of its execution.

The sociopolitical and cultural contexts of the Transvaal gold rush are handled well in Frank Welsh's (2000) excellent revised edition of *A History of South Africa*. Detailed documentation of the miners' lifeways is lacking, but, as should be expected because of its economic significance, there is a vast literature on the geology of South Africa. Much of this information appears in the technical reports of mining companies and is not cited here. Lime was not only used in the Macarthur-Forrest cyanide process but was also a constituent of toothpaste and cement. In addition, it served as flocculate in sugar refining and water purification and as an essential chemical agent in the production of commercial uranium, steel, and paper (Douglas, 1969).

### The Deep Womb of Humanity

The second edition of Stanley's (1999) *Earth System History* is a good primer on historical geology that includes discussion of the evolution of the Earth's crust and its continents. For more specific information, a good overview of cave formation in the Witwatersrand is found in Brain (1981a). Brain (1981a) is also the authoritative work on bone-accumulating process in southern African caves; see also Andrews (1983); Pickering (1999); Pickering, Clarke, and Moggi-Cecchi (2003). Hodgins, Brook, and Marais (2007) describe modern baboon mummies deposited in a Namibian cave by a death trap process. Vrba (1981), Cooke (1991), Pickering, Clarke, and Heaton (2004), Mokokowe (2005), and Kibii (2007) discussed various fossil death trap assemblages from the Witwatersrand caves.

*The Family Tree*

Taung, like most other South African paleoanthropological sites, was discovered as a result of the economic pursuits of those indefatigable, early twentieth-century lime miners. It, however, formed in a different manner than did the caves of the Witwatersrand. Taung is, instead, a tufa cave, which evolve in dolomite escarpments. Ground water collects in joints and other gaps in the dolomite blocks that comprise an escarpment. The water, now charged with dissolved lime, eventually empties across the escarpment in the form of springs. The discharge from those multiple springs can coalesce within a streambed or, if the escarpment is of shallower topography, fan out across it. Even in the former case, water flow slows at the edges of any shallow streambed. It is in these trickling portions of the flow that algae and moss take hold. When those organisms die and bake in the sun, their calcium carbonate skeletons remain and accumulate, stacked ultimately, over great time spans, into thick deposits of lime. These layers of lime are called "tufa flows" and often extend over the edges of the escarpments upon which they form. Such overhangs, like the underground caves of the Witwatersrand, provided shelter for bone-accumulating animals in the past. In the case of Taung, the bony residues left behind by the activities of prehistoric animals mixed with distinctive red sands blown in from the Kalahari Desert. These materials were eventually sealed from the outside world by the continued, slow, and steady accretion of tufa across the opening—that is, until the lime miners discovered the cave (see McKee [1993] for a good summary of tufa cave formation).

Goodman (1963, 1998) provides informative reviews of biochemical relatedness in apes and humans and its relevance to dates of divergence between the lineages; see also Ruvolo (1997) for evidence from molecular studies and Bradley (2008) for evidence from genomic studies. Fabre, Rodrigues, and Douzery (2009) present a good review of the pattern of primate evolution as based on DNA evidence. An important paper by White et al. (2006) discusses the paleontological evidence for the emergence of *Australopithecus* and its impact on interpretations of the tempo and mode of early hominin evolution.

Dart's (1925) initial description of the Taung Child appeared in the journal *Nature*. I choose here to cull only a few quintessential publications from the vast literature on putative late Miocene and early Pliocene hominins and on the various species of the genus *Australopithecus*: for *Ardipithecus kadabba* see Haile-Selassie and WoldeGabriel (2009); for *Ardipithecus ramidus* see White et al. (2009a) (for information on *Ardipithecus ramidus* from Gona, see Semaw et al. [2005]); for *Sahelanthropus* see Brunet et al. (2002); for *Orrorin* see Senut et al. (2001); for *Australopithecus anamensis* see Ward et al. (1999); for *Australopithecus afarensis* see Johanson et al. (1982); for *Australopithecus africanus* see Broom and Schepers (1946); for *Australopithecus robustus* see Broom and Robinson (1952); for *Australopithecus aethiopicus* see Walker et al. (1986); for *Australopithecus boisei* see Tobias (1967); for *Australopithecus garhi* see Asfaw et al. (1999); for *Australopithecus prometheus*, represented by the Sterkfontein Cave "Little Foot" skeleton, see Clarke (1998, 1999, 2002, 2007, 2008, 2013); see also Clarke and Tobias (1995); Pickering, Clarke, and Heaton (2004). Wood and Lonergan (2008) have produced an up-to-date and clear-cut compendium of the morphology of these early hominin forms, excluding the newly published information on *Ardipithecus ramidus* gleaned from the analyses of Ardi's skeleton. A series of articles on that skeleton by White et al. (2009a,b), WoldeGabriel et al. (2009), Louchart et al. (2009), Suwa et al. (2009a,b), Lovejoy (2009), and Lovejoy et al. (2009a,b,c,d) appears in a special issue of the journal *Science*. The description of the strange foot of *Ardipithecus ramidus* joins the ranks of many new revelations about the complexity of this anatomical element in our early bipedal ancestors: as discussed in the text, Clarke and Tobias (1995) conclude that "Little Foot" possessed an opposable big toe; *Australopithecus afarensis* is interpreted as lacking a grasping big toe and having a typical modern human longitudinal arch in its foot (Latimer and Lovejoy, 1982, 1990a,b; Ward, Kimbel, and Johanson, 2011); the newly described *Australopithecus sediba*, from South Africa, has a foot with a developed arch, sharply angled, apelike heel, and highly mobile ankle joint (Zipfel et al., 2011); most recently, Haile-Selassie et al. (2012) describe a hominin partial foot skeleton from Burtele, Ethiopia, that has an opposable big toe (DeSilva, Proctor, and Zipfel [2012] describe a hominin foot bone from a 2 million-plus-year-old layer at Sterkfontein Cave as similar to the same bone in the Burtele foot). The apelike Burtele foot remains are, at 3.4 million years old, contemporary with *Australopithecus afarensis*, which, as summarized previously, had a demonstrably humanlike foot; the coexistence of these two distinct types of hominin feet indicate, in turn, that two separate species of ape-man lived during the time of Lucy. The Laeotoli ape-man footprint trails were announced, analyzed, and debated in Leakey and Hay (1979), White (1980), Leakey (1981), Stern and Susman (1983), White and Suwa (1987), Tuttle (1987), Tuttle et al. (1991), Raichlen et al. (2008, 2010). Haile-Selassie, Suwa, and White (2004) report on their investigation of the primate upper canine tooth honing complex and its modification in *Ardipithecus*. Kimbel et al. (1996) and Kimbel, Johanson, and Rak (1997) describe the Hadar upper jawbone fossil and attribute it to the genus *Homo;* see also a more recent discussion of the upper jawbone of early *Homo* in Blumenschine et al. (2003) and Clarke (2012). Reader (1981), Lewin (1997), and Gibbons (2007) provide entertaining histories of the search for early hominin fossils and competing ideas of human evolution.

Estimates of body size play an important role in reconstructing early hominin social organization and behavior: see, for example, Clutton-Brock (1985) and Leigh (1997). Living primates with considerable disparities in the sizes of males and females, like gorillas, are usually characterized as having high degrees of antagonistic male–male competition for mating access to females. The opposite is true in species, such as modern humans, in which males and females are closer in body size. Estimates of humanlike intersexual differences in *Australopithecus afarensis* males and females are from Reno et al. (2003, 2010), who used the dimensions of the heads of the thigh bones of Lucy and other individuals to extrapolate live body size.

In contrast, Lockwood et al. (2007) analyzed a large collection of adult *Australopithecus robustus* crania (we know they are adults because each has erupted wisdom teeth) from the South African cave sites of Swartkrans and Drimolen. Using wear on the chewing surfaces of the teeth as a proxy,

Lockwood and his coworkers assessed the relative ages of the crania. Small, lightly built crania of every adult age are represented—from the youngest to among the oldest—and these are all inferred to be females. It is assumed, by extension, that females must have reached their full skeletal development (that is, they stopped growing) by the time their wisdom teeth erupted. The largest crania, however, are always old. Stated another way, there are no large, young crania. Those large, old adult crania also show the most extreme robusticity and skeletal ruggedness of the sample, so it is inferred for good reason that they are all male. This leaves small, young crania without the delicate features of females. These are inferred to be males, who, although adult with erupted wisdom teeth, did not survive long enough to grow to complete maturity and the full physical bulk that was achieved by old males. Lockwood et al. (2007) used these observations to argue for social organization in *Australopithecus robustus* that was decidedly different than that reconstructed by Lovejoy (2009) for earlier *Ardipithecus* (see discussion in text of chapter 2) and monomorphic (that is, little difference in male and female body sizes) *Australopithecus afarensis,* saying that

> [extended] male growth occurs in primates when male reproductive success is concentrated in a period of dominance resulting from intense male-male competition. . . . Climbing the dominance hierarchy typically involves not only an increase in size but also changes in soft-tissue anatomy and, in some taxa, coloration. . . . Bimaturism [males and females maturing at different rates and in different patterns] can be interpreted as part of a male-strategy to delay competition with high-ranking males until the likelihood of success is greatest.

An enormous *Australopithecus robustus* molar, the largest on record, is known from the site of Gondolin (Menter et al., 1999). Without this single fossil we would not fully appreciate the extremeness of the difference in the size of male and female *Australopithecus robustus*. Based on taphonomic grounds, Grine et al. (2012) argue that the huge Gondolin molar is *"expectedly* rare" (my emphasis); they posit that, unlike smaller females and juveniles, the size of *most* very large males (Gondolin excepted) ensured that they were rarely killed, eaten and deposited in caves. Studies of *Australopithecus robustus* hip and leg bones corroborate that the species was highly sexually dimorphic (Susman, 1988; Susman, de Ruiter, and Brain, 2001; Ruff, 2010; Pickering et al., 2012).

### Precocious Beasts Who Flouted Sacred Cows

The reason that the skulls of the three robust ape-man species are so similar is because they all have small front teeth and huge back teeth, a unique configuration that (regardless of species) "forced" the other parts of their skulls to grow in the same ways (McCollum, 1999). The back of the lower jaw had to become taller to support their massive molars, which, in turn, required thick jaw muscles and heavily buttressed faces and braincases onto which those muscles could attach. The small front teeth affected the shape of the floor of nose in which they were anchored. The palate is a developmental junction between the mouth and the nose and, in the robust ape-men, grew very thick in order to accommodate the differential growth patterns between these other two areas of the skull. An edited volume by Fred Grine (1988) deals exclusively with the robust australopithecines, containing excellent chapters on the bony anatomy of that group and hypothetical considerations of its evolutionary relationships. (More recent reviews of *Australopithecus boisei,* the East African robust ape-man, are by Wood and Strait [2004], Wood and Constantino [2009], and Constantino and Wood [2009].) Grine, Fleagle, and Leakey (2009) provide a complementary volume concerning the emergence of the genus *Homo.*

Semaw et al. (1997) report on the earliest known stone tools from 2.6 million-year-old deposits at Gona, in Ethiopia. De Heinzelin et al. (1999) make a circumstantial case for *Australopithecus garhi* having used stone tools to butcher animal carcasses 2.5 million years ago (see also Domínguez-Rodrigo et al. [2005] for evidence of 2.6 million-year-old butchery from Gona), which joins a morphological analysis of the species that concludes it might well be the direct ancestor of the genus *Homo* (Asfaw et al., 1999). Strait and Grine (1999, 2001; see also Kimbel, 2009) dissent in their opinion that nothing in the morphology of *Australopithecus garhi* indicates it was directly ancestral to the genus *Homo.* Recently, Berger et al. (2010; see also Dirks et al., 2010; Pickering et al., 2011) announced *Australopithecus sediba,* a morphologically derived, 1.98 million-year-old species of ape-man, and proposed that it might be the direct australopithecine ancestor of the genus *Homo.* This opinion is contended; *Australopithecus sediba* might, instead, simply be a time-successive chronospecies of the earlier occurring South African species *Australopithecus africanus.*

It is best to think of two "savanna hypotheses" in paleoanthropology. One is applied to a more recent stage of human evolution to explain the divergence of lineages that produced, respectively, early *Homo* and the robust australopithecines; the other is the classic application of a "savanna hypothesis" to explain the emergence of the hominins, much earlier in time, between 8 and 4 million years ago. I owe Manuel Domínguez-Rodrigo a great debt for correcting my stereotypic understanding of the development of the classic "savanna hypothesis" of hominization. Darwin's (1871) oblique reference to ecological change, protohuman descent from trees, and the adoption of bipedalism is found in his classic work *The Descent of Man and Selection in Relation to Sex.* Leakey (1934) and Dart (1957) mentioned the idea of the hominin–savanna evolutionary link only later. The association of the earliest purported hominin genus, *Ardipithecus,* with the fossils of woodland plants and animals (as well as the isotopic signatures of those fossils) in the Middle Awash of Ethiopia, is argued to falsify the savanna hypothesis of human origins (see WoldeGabriel et al. 1994, 2009; White et al., 2009a,b; Louchart et al., 2009), but see also Semaw et al. (2005) and Levin et al. (2008) for reconstructions of more open *Ardipithecus* habitat at Gona, Ethiopia. A recent study demonstrated that carbon isotopes in soils can be used to quantify the fraction of woody cover in tropical ecosystems (Cerling et al., 2011). The researchers then applied the same analyses to ancient "fossilized" soils (called paleosols) recovered from important hominin sites throughout East Africa. Their results indicate the "prevalence of open environments at the majority of hominin fossil sites in eastern Africa over the past 6 million years" and "leave the savannah hypothesis as a viable scenario for

explaining the context of earliest bipeldism, as well as potentially later evolutionary innovations within the hominin clade." Pickering and Domínguez-Rodrigo
(2012) discuss reconstructions of Miocene hominin paleohabitat in some detail.

The modern scientific literature on the relationship of climatic change and
human evolution is storied and extensive, beginning with the work of Bob Brain
(1981b, 1984). Since that time, refinement and testing of hypotheses that link
climate change and evolution have benefited from an ever-increasing fossil
record, advances in the study of the deep-sea paleoclimate record (which can be
related to the terrestrial record), and a much greater appreciation of the impact
of tectonic and orbital forcing on the paleoenvironment of Africa throughout
the course of hominin evolution. Three essential relevant works include Vrba
et al. (1995); Bobe et al. (2007); Christensen and Maslin (2007).

Very influential on mid-twentieth century paleoanthropology was the
opinion of the preeminent biologist Ernst Mayr (1950), who argued that the
ecological principle of competitive exclusion should be applied to human
evolution—meaning that only one species of hominin could exist at any given
time. Adherents of Mayr's view saw hominins with robust skulls as males and
those with lightly built skulls as females of the same species.

John Robinson skirted the roadblock of competitive exclusion by arguing
that more than one species of hominin could coexist if they were separated
ecologically, relying on different foods and different dietary strategies. (The
idea of competitive exclusion in human evolution, formalized as the "single
species hypothesis," became a persistent minority anthropological opinion [e.g.,
Brace, 1967; Wolpoff, 1968, 1971, 1976, 1978] in due course.) Robinson wrote
prolifically about his "dietary hypothesis," which divided different forms of
contemporaneous hominins into specialized herbivores (robust ape-men) and
generalized omnivores (gracile ape-men and early *Homo*); good distillations of
the idea are found in Robinson (1962, 1963). For data on the nutritional and
caloric content of savanna nuts and roots see Murray et al. (2001) and
Schoeninger et al. (2001). For up-to-date reviews on the voluminous
studies of early hominin microscopic dental wear and isotopic signatures, as
well as general considerations of hominin diets, see, e.g., Scott et al. (2005);
Lee-Thorp and Sponheimer (2006); Ungar, Grine, and Teaford (2006); Lucas
et al. (2008); Ungar (2011); Ungar and Sponheimer (2011); Ungar et al. (2012).
Matt Sponheimer's revelatory research on stable isotopes includes, among
many other publications, Sponheimer and Lee-Thorp (1999) and Sponheimer
et al. (2006). Details of the *Australopithecus africanus* brucellosis study are
found in D'Anastasio et al. (2009). C. K. Brain (Brain et al., 1988; Brain and
Shipman, 1993) noticed remarkable similarities between the wear on the heads
of screwdrivers he used to excavate in coarse-grained sediments at Swartkrans
Cave and the wear on fossil bone tools that he recovered in those excavations.
Prompted by this congruency, Brain conducted a series of experiments using
modern bones to dig for edible resources that would have also been available
to early hominins armed with similar bone tools. Brain based the "root-digging
hypothesis" for robust ape-men on his experimental findings. Backwell
and d'Errico (2001) presented the competing "termite-digging hypothesis"
based on their subsequent analysis of the Swartkrans bone tools and their own

experimental program. Pickering (2006; see also Pickering and Heaton, 2009; Wood and Strait, 2004) presents a synopsis and critical review of the "dietary hypothesis" and current research on South African early hominin diets.

### Survival Mode

The seeming evolutionary contradiction posed by an organism that typically behaves as an ecological generalist but who also displays extreme morphological specializations is referred to as "Liem's Paradox" (see Liem [1980] for articulation of the paradox as it applies to cichlid fishes). In paleoanthropology, the robust australopithecines—with highly specialized skulls and teeth and a diet that was probably generalized—are the prime example of Liem's Paradox; see especially the recent, direct statement on this topic from Ungar, Grine, and Teaford (2008). In a proposed solution to the paradox, Laden and Wrangham (2005) hypothesize a gorilla-like strategy for robust australopithecines as it relates to their exploitation of fallback foods. Marshall and Wrangham (2007) and Constantino and Wright (2009) are recent discussions of the importance and consequences of fallback foods in primate evolution.

Dominy et al. (2008) present findings relevant to the "$\delta^{12}C/\delta^{13}C$ *Australopithecus* conundrum" in the journal *Evolutionary Biology*. Lucas et al. (2008; see also Lucas, 2004) review the use of dental enamel to indicate diet in mammals and forward the hypothesis that much of the microscopic wear on the chewing surfaces of teeth is created by nondietary grit that is consumed incidentally with certain foods. (Another drawback of reliance on the microscopic wear on teeth to inform about diet is that the wear from any individual meal is rapidly obliterated by wear from subsequent meals [Grine and Kay, 1988; Teaford and Oven, 1989]; this results in a "last supper" effect that may mask important elements of an extinct animal's diet. Likewise, Sponheimer and Lee-Thorp [1999; see also Yeakel et al., 2007] caution that the isotopic signal of an important but rarely consumed fallback food can be overwhelmed by the signal from foods that are eaten more regularly.)

Some important studies of the face of *Australopithecus* include those of Broom and Robinson (1952), Tobias (1967), Rak (1983), Wood (1991), Suwa et al. (1997), Lockwood (1999), Lockwood and Tobias (1999, 2002), Keyser (2000), Kimbel, Rak, and Johanson (2004) and Strait et al. (2009); the quote on pages 36–37 is from the latter reference. Strait et al. (2009) also conclude that

> key aspects of [australopithecine] craniofacial morphology are ... likely to be [primarily] related to the ingestion and initial preparation of large, mechanically protected food objects like large nuts and seeds. These foods may have broadened the diet of these hominins, possibly by being critical resources that [australopithecines] relied on during periods when their preferred dietary items were in short supply. Our analysis reconciles apparent discrepancies between dietary reconstructions based on biomechanics, tooth morphology, and dental microwear.

Following earlier observations of chips in the enamel of fossil hominin teeth (e.g., Robinson, 1954; Wallace, 1973), Constantino et al.'s (2010) updated analysis of such damage confirms that hominins probably ate large, hard objects, like seeds and nuts. Collectively, the recent results on early hominin diet indicate that most ape-men were fairly generalized feeders, unconstrained by a highly specialized diet, and able to subsist on low quality foods in hard times. Wood and Strait (2004) and Pickering (2006) concluded much the same in their earlier reviews of evidence for early hominin dietary and behavioral strategies. For the exception, startling new isotopic results suggest that *Australopithecus boisei* was a specialized grass eater (van der Merwe et al., 2008; Cerling et al., 2011b). The microscopic wear of *Australopithecus boisei* teeth seems to support (or, at least not contradict) the isotopic conclusions (Ungar, Grine, and Teaford, 2008; Ungar et al., 2012), but, complicating the picture, some teeth of *Australopithecus boisei* have chips in their enamel indicating their forceful biting of large, hard food objects (Constantino et al., 2010). The quote on page 37 is from Sponheimer et al. (2006).

Finally, a new study of the diet of *Australopithecus sediba*, from the ~2 million-year-old South African cave site of Malapa, combines results on microscopic wear, isotopic signal and phytolith analysis (see chapter 4 and its notes for discussion of the analysis of phytoliths recovered on ancient stone tools at the site of Peninj, in Tanzania) (Henry et al., 2012). The latter line of evidence is the most exciting as the *Australopithecus sediba* study is the first time that phytoliths (silica particles derived from plant cells) have been extracted from the dental calculus (or hardened plaque) of a fossil hominin. The study thus demonstrates a new approach to reconstructing early hominin subsistence in contexts in which fossil teeth preserve tartar and its potential dietary inclusions.

### Man, Nasty and Brutish

The story of the Piltdown Man hoax is well reported in various accounts. Of the concise variety, Lewin's (1997) is among the best in contextualizing its impact on the rejection of the Taung Child as a legitimate early hominin; longer accounts include those of Spencer (1990) and Walsh (1996). A thriving subgenre deals with the whodunit aspect of the fraud. Even Sir Arthur Conan Doyle, the creator of Sherlock Holmes, has been offered up as suspect among the slew of suggested perpetrators. But recently, Russell (2003) documents a disquieting pattern of archaeological fraudulence by amateur antiquarian Charles Dawson in the twenty years leading up to his "discovery" of the first Piltdown Man remains. Dawson was an early suspect in the crime, and Russell's careful documentation of Dawson's seemingly inborn duplicity goes a long way in justifying those suspicions. Gould (1980, 1981) and Thackeray (1992, 2009, 2011, 2012) judge the motives of the Piltdown Man forger, who they believe was Father Pierre Teilhard de Chardin (a Jesuit priest, paleontologist, and discoverer of Piltdown Man's canine tooth) more charitably (see also Harrison Matthews, 1981; Gardiner, 2003). Gould and Thackeray argue that the fraud was probably hatched by Teilhard (possibly acting in concert with Dawson and Martin Hinton, a scientist at the British Natural History Museum), starting as a harmless joke but then unraveling into a full-blown hoax when, because the "fossil" was accepted as genuine by so many experts, Teilhard and his co-conspirators could now never reveal the truth without damaging their careers.

Solly Zuckerman's autobiography was published in two installments, ten years apart, in 1978 and 1988; Peyton (2001) has produced a more recent biography of Zuckerman. Zuckerman's attack against the legitimacy of australopithecines as hominins (see representative distillations in Zuckerman, 1950, 1954) was disregarded widely in paleoanthropology by the end of the 1950s, but Charles Oxnard (1975, 1984), a vocal disciple of multivariate morphometrics, kept up the drumbeat well into the 1980s. The most current (and lively) voice of dissent against the hominin-status of *Australopithecus* is Estaban Sarmiento's (e.g., 1987, 1995, 1998, 2000). Consider a choice passage from Sarmineto (2000: 15):

> [The] dogma Dart and Le Gros Clark ... dreamed-up fifty years ago to convince the world that australopithecines were human ancestors ... relied on faulty reasoning, postulating that all distinguishing human characters present in australopithecines are derived and correlates of human bipedalism and tool use.... Interpretations of australopithecines as terrestrial bipedal hominids are no longer tenable with the realization that reduced canines, nonsectorial [third premolars], and many alleged australopithecine bipedal characters are primitive for humans and African apes, if not hominoids.... Alternative interpretations ... indicate that australopithecines practiced both terrestrial quadrupedal and bipedal behaviors and, if arboreal, were restricted to relatively large-diameter supports. Their behaviors and associated anatomy, therefore, are as likely to correspond with human, gorilla, or chimpanzee ancestors, or the most recent human or African ape ancestor.... If Plio-Pleistocene fossils are all hominids (in the classical sense) where are the African ape lineages?

I consulted the standard biography of Robert Broom, by Findlay (1972), as well as Broom's (1950) own *Finding the Missing Link*. Dart's first impression of Broom, quoted on page 42, is from Dart and Craig (1959). Broom's last completed monograph, coauthored with John Robinson, is *Swartkrans Ape-Man Paranthropus crassidens* (1952). Broom's (1932) early work on South Africa's mammal-like reptiles has obviously been surpassed in the past eighty years (for an overview, see McCarthy and Rubidge, 2006), but his research on the topic still stands as quite historically important.

New South African ape-man fossils were unearthed at a relatively rapid clip following the announcement of the Taung Child. The history of this remarkable phase of discovery is convoluted by publications of the period that announced new finds with slews of supposedly unique genera and species, which are now ultimately sunk into just one or two genera and only two species. Briefly, Broom (1936) named the first adult australopithecine fossil from Sterkfontein *Australopithecus transvaalensis,* distinguishing it initially at the species level from the Taung Child (*Australopithecus africanus*). Not content with the specific distinction between *Australopithecus africanus* and *Australopithecus transvaalensis,* Broom (1938) eventually went further, moving the species *transvaalensis* into its own genus, *Plesianthropus*. Broom (1938) also described the first known robust australopithecines in the same article, applying the name *Paranthropus robustus* to his new fossils from the site of Kromdraai. Further, Broom (1949) considered the Kromdraai robust ape-men sufficiently different from the robust hominin remains recovered at Swartkrans to place the latter in its own species, *Paranthropus crassidens*. A final complication came from Dart's (1948) placement of gracile australopithecine remains from Makapansgat into the novel taxon, *Australopithecus prometheus,* the species name *prometheus* a taxonomic nod to his contention that this hominin controlled fire.

Today, the taxonomic picture for South African hominins is comparatively uncomplicated: most experts now consider *Australopithecus africanus, Plesianthropus transvaalensis,* and *Australopithecus prometheus* to belong to the same species, *Australopithecus africanus.* (However, Ron Clarke [2012] has recently revived *Australopithecus prometheus* to accommodate a group of fossils from Makapansgat and Sterkfontein, which he believes are contemporary with but taxonomically distinct from *Australopithecus africanus* [see discussion in chapter 5 text].) Likewise, the current majority view sees only one species of robust ape-man in South Africa, *Paranthropus robustus,* which encompasses both the Kromdraai and Swartkrans fossils as well as more recently discovered materials from Coopers, Sterkfontein, Drimolen, and Gondolin caves. As is advocated tacitly here, some specialists lump to even greater extent, placing the South African robust australopithecines into the same genus as the other ape-men by calling them *Australopithecus robustus. Homo erectus* remains from Swartkrans were first called *Telanthropus capensis* (Broom and Robinson, 1949, 1950), their affinity to the genus *Homo* accepted only later (e.g., Robinson, 1961). Many contemporary experts assign early African *Homo erectus* remains to the species *Homo ergaster* or *Homo leakeyi* (see, e.g., Clarke 1994a; Wood and Collard, 1999). Taxonomies that view *Homo ergaster/leakeyi* as a legitimate species usually relegate the name *Homo erectus* to indicate a specially derived, Asian side branch of human evolution, with a tangential rather than direct biological connection to modern people.

Dart's literary output on the "killer ape" and "osteodontokeratic hypotheses" was inexhaustible between 1949 and 1965. Five representative papers, of the nearly thirty from this time period, include Dart (1949, 1953, 1956a, 1957, 1960). The quote on page 45 is from Dart (1953). Robert Ardery popularized Dart's ideas in the books *African Genesis* (1961), *The Social Contract* (1970), *The Territorial Imperative* (1971), and *The Hunting Hypothesis* (1976).

### CHAPTER 3: TAMPING THE SIMIAN URGE

Matt Cartmill's 1993 book, *A View to a Death in the Morning,* explores ideas of human hunting from myriad perspectives, including the scientific and the popular, and across time, starting with the ancient Greeks up to modern day. Among Charles Morris's works on aggressive early humans is his bluntly titled 1890 *American Naturalist* article, "From Brute to Man." Harry Campbell (e.g., 1904, 1917, 1921) published his views on prehistoric violence in a series of articles in the British medical journal, the *Lancet;* Carveth Read expanded his similar views to book length in 1925's *The Origin of Man.*

Along with Niko Tinbergen, Konrad Lorenz founded the science of ethology, the study of animal behavior. Burkhardt (2005) has written a recent account of Tinbergen and Lorenz's new science that also deals to an extent with the sociopolitical implications of their work. Kalikow (1983), Deichmann (1999), and

Foger and Taschwer (2001) are more focused studies of Lorenz's ties to Nazism. Lorenz's own writings are readable and provocative: for examples, see *King Solomon's Ring* (1952), *On Aggression* (1966), *Civilized Man's Eight Deadly Sins* (1974), *The Year of the Greylag Goose* (1979), and *The Foundations of Ethology* (1981). See Friedman (2001) for a recent history of and references relevant to Erik Erikson's development of the concept of "pseudo-speciation."

An edited volume by Boesch et al. (2001) compares and contrasts the behavior of common chimpanzees and bonobos. *Bonobo: The Forgotten Ape* (de Waal and Lanting, 1997) is a sympathetic look at this lesser-known species of the genus *Pan*. See also Kano (1982, 1992) and Furuichi (2011) for more on bonobo socioecology. Savage-Rumbaugh and Lewin (1994) chronicle the life and cognitive achievements of Kanzi, a captive bonobo with remarkable abilities to understand spoken English and to communicate with humans using a sophisticated signboard. Kanzi has also been taught to produce simple stone tools, like those of prehistoric hominins, and to use them to cut bindings tied around treat boxes (Toth et al., 1993; Schick et al., 1999). Stanford (1998b) contends that many of the proposed behavioral discrepancies between common chimpanzees and bonobos may actually have more to do with the relative paucity of observational data on bonobos than any real differences between the species. The quote on page 48 is from Wrangham and Peterson (1996). The one million-year-ago divergence estimate for chimpanzees and bonobos is based on the work of Becquet and Przeworski (2007) and Hey (2010).

*Science's Cover Girl and Her Dark-Hearted Wards*

There are several biographies of Louis Leakey, including *White African,* his 1937 memoir. Much happened in Leakey's life (1903–72) subsequent to that early date, so consult Morell's (1995) definitive biography of the whole Leakey family, including Louis's archaeologically accomplished wife, Mary, and paleontologically accomplished son, Richard, and daughter-in-law, Meave. Mary (1984) and Richard (1983) also published autobiographies.

Dian Fossey's pioneering research on mountain gorillas is best known from her 1983 book *Gorillas in the Mist*. Biruté Galdikas published her autobiography in 1996. Many of Jane Goodall's groundbreaking contributions to our understanding of chimpanzee behavior are summarized in her 1986 monograph, *The Chimpanzees of Gombe,* and in the more introspective *In the Shadow of Man* (1971), *My Life with Chimpanzees* (1996 [source of the quote on page 50]), and *Through a Window* (2000). Peterson's (2008) biography of Goodall is also a good resource.

*Demonic Males: Apes and the Origins of Human Violence,* by Richard Wrangham and Dale Peterson (1996), is a wonderful provocation that explores the links between primate maleness and primate viciousness. The book deals with all the major topics discussed in this section of chapter 3; it is also the source of the quote on page 51. Also see Wrangham (1999 [source of the quote on page 52]) for a good review of lethal raiding and coalitionary killing, as well as Wilson and Wrangham (2003) and Wrangham et al. (2006). More specific case studies of wild chimpanzee intergroup violence and discussions of how they relate to increased access to food and females include: Nishida et al. (1985); Goodall (1986); Watts (2004); Williams et al. (2004); Hashimoto and Furuichi (2005).

### The Triumph of Self-Possession

Female exogamy and male philopatry are traits shared by chimpanzees, bonobos, and human hunter-gatherers (Ember, 1978; Ghiglieri, 1987; Wrangham, 1987). Unusual among other primates, some experts hypothesize that these conditions also characterized the most recent common ancestor of all three species. The principle of inclusive fitness recognizes that an individual's genetic success is increased not only by the number of viable offspring he or she produces but also by supporting the success of his or her "genetic equivalents" (that is, his or her other relatives). The biologist W. D. Hamilton (1964) argued that inclusive fitness provides an evolutionary explanation for altruism in some group-living organisms.

Guthrie's (2005) book is ostensibly his extended thesis that art functioned as outlet for youthful machismo in the Paleolithic, but it is also filled with affectionately told hunting anecdotes, both personal and those collected from various traditional cultures. Other memoirs of hunter-gatherer life, less formal than strictly academic works, include Marshall Thomas's (1959) *The Harmless People,* van der Post's (1980) *The Heart of the Hunter,* Shostak's (2000) *Nisa,* and Stephenson's (2000) *Language of the Land;* these references provide a good feel for the human forager lifestyle in tropical and subtropical Africa.

Gilby et al. (2010) contend that blunt meat-for-sex exchanges are not only rare among humans, but that, even among wild chimpanzees, they "are so rare, and so different in nature from exchanges among humans, that with respect to chimpanzees, sexual bartering in humans should be regarded as a derived trait with no known antecedents in the behavior of wild chimpanzees."

See Hawkes et al. (2001) for data showing that the best Hadza hunters have the best (in an evolutionary sense) wives. Important book-length works on the Hadza include Woodburn's (1970; see also Woodburn, 1968) classic material culture and ethnographic study, and Marlowe's (2010) recent research on Hadza culture and evolutionary ecology.

An interesting study of how Hadza rank various foods—including honey and meat as, respectively, ranks one and two—is found in Marlowe and Berbesque (2009). Like hunting, honey foraging is nearly exclusively a men's activity in most modern foraging groups, such as the Hadza. Compared to plant foraging, traditional hunting and honey gathering are both relatively risky pursuits. The dangerous aspects of hunting are discussed in the text, whereas honey gathering not only exposes foragers to stinging bees, but, when hives are located high in trees, it requires a significant degree of climbing ability, strength, stamina, and athleticism. African killer bees produce the honey that is most prized by the Hadza, and their combs also often contain fat, tasty larvae. The problem is that African killer bee hives are usually located high in baobab trees, which can easily be more than 30 or 40 feet in height. That means a Hadza man needs to climb the baobab in order to get the honey and larvae, but baobabs don't have low limbs. The man solves this problem by cutting a bunch of short hardwood pegs, each of which he sharpens on one end. Then, the man shoves all but one of the pegs into his belt or waistband. He uses the blunt back of his metal ax blade to hammer the remaining peg into the baobab's soft bark, high above his head, and then drives another one shoulder-width apart from the first. The man pulls himself up on these pegs and proceeds up the baobab trunk, driving peg after peg above his head until he reaches the hole leading to the hive. Yes, this *is* as perilous as it sounds: in order to hammer in each consecutive peg, the man, each time, must balance on the foot-pegs and hold a hand-peg with only a single hand as he uses the other hand to swing his ax above his head in order to drive the next peg. Once the man reaches the position of the hive, he usually has to widen the hole by cutting it open with his ax—all the while balanced precariously, maybe 30 or more feet above the ground, on the pegs he has set in the tree. With the hole opened, the man can reach in and start extracting the comb (which he hopes is laden with honey and larvae); the man often uses smoke from a smoldering torch (which he has also carried up the tree with him!) to quell the angry bees, but this is never 100 percent effective, and most guys still get quite stung up. Finally, the man needs to load the bee product into a container and lower it, and himself, down the tree. This whole procedure is doubly fascinating when one realizes that birds, not Hadza, are often the discoverers of the hives. Honeyguide birds and the Hadza have formed a symbiotic relationship in which they communicate with a series of complicated whistles, a bird eventually leading a man to a hive. The honeyguide cannot, however, open up holes leading to hives, so the Hadza man will leave part of the captured honeycomb and its grubs to the bird helper, rewarding its assistance and consolidating the collaborative relationship between human and animal. African honeyguide birds are discussed in Short et al. (2002). Hadza tuber foraging is discussed in Vincent (1985), Schoeninger et al. (2001), and Marlowe (2010).

The John Hawks quotes on page 55 are taken from his blog (johnhawks. net), which is dedicated to all issues paleoanthropological. The *MYH16* story, summarized, is that the gene is

> specifically expressed in the jaw muscles of humans and monkeys. But, surprisingly, a mutation in the human gene prevents the accumulation of *MYH16* protein. Stedman et al. [2004] found that, by contrast, all non-human primates for which genome sequence could be obtained have an intact copy of the gene, and have a high level of *MYH16* protein in their jaw muscles. An analysis of the time at which the mutation arose during hominid evolution places it at about 2.4 million years ago, the period just before the evolution of the modern hominid cranial form. These findings suggest a seductive hypothesis: that a decrease in jaw-muscle size, produced by inactivation of *MYH16*, removed a barrier to the remodelling of the hominid cranium which consequently allowed an increase in the size of the brain. (Currie, 2004)

Many experts are dubious about the causal link proposed between large skull musculature and the inhibition of brain growth (e.g., McCollum et al. 2006). Walker (2009 [source of the quotes on pages 55 and 56]) presents his hypothesis of great ape strength and human muscular speed in the journal *Current Anthropology*.

Gonzales's (2004) *Deep Survival* is highly recommended. It is an exhilarating read and makes a compelling case for core personality traits and emotional assets that separate survivors from victims.

### The Methodical Apes

There is large literature on chimpanzee hunting, including important primary works by Takahata et al. (1984), Boesch and Boesch (1989), Stanford (1998a), Stanford et al. (1994), Uehara (1997), Mitani and Watts (2001), and Watts and Mitani (2002). Newton-Fisher (2007) provides a recent summary of previous work on the topic, and Stanford's (1999) *The Hunting Apes* is a good synopsis of comparative research on hominoid hunting written for general readership. Pruetz and Bertaloni (2007) describe the unique spear hunting of the Fongoli chimpanzees. As an aside, I am proud to have been part of the research team that saw to falsifying another long-presumed distinction between humans and chimpanzees. Until our observations and descriptions of wooden digging tools from the chimpanzee field site of Ugalla, in Tanzania, chimpanzees were not believed to dig up edible tubers using tools, as is commonly done by modern human foragers in subtropical Africa (Hernandez-Aguilar, Moore, and Pickering, 2007).

Examples of the fascinating comparative work on humans and dogs that demonstrates similarities in their cooperative communication abilities include: Hare and Tomasello (2004, 2004), Hare et al. (2002), and Tomasello et al. (2003). Summary of that research appears in Hare (2007), which is also the source for the quote on pages 59–60. More recent work, building on that of Hare, Tomasello, and their colleagues, demonstrates another striking correspondence in the social-cognitive skills of humans and dogs. Topál et al. (2009) showed experimentally that ten-month-old humans search relentlessly for a hidden object where they were first shown it to be hidden, even after the children *observed* it being hidden elsewhere. Domestic dogs make the same mistake. In other words, young children and dogs trust more what they have learned previously from an adult human than they do contradictory information gathered with their own eyes. Dogs do, however, trust their own eyes when a different adult human than the original hider places the object of desire in a new location. In contrast, children do not modify their faulty search efforts when a new adult is introduced into the experiment. The difference is potentially quite important, as Topál et al. suggest that it means that human children are responsive to true generalized cultural learning, while dogs lack such a capability.

There is considerable disagreement about when dogs were first domesticated. Various genetic studies provide an unsatisfying range for their point of origin between 15,000 and 100,000 years ago (Wayne et al., 1997; Vilà et al., 1997; Savolainen et al., 2002; Bardeleben et al., 2005; Lindblad-Toh et al., 2005). The earliest archaeological claim for domesticated dogs is from the 31,700-year-old site of Goyet, in Belgium, where the skull measurements of several large canid fossils are clearly distinct from those of recent wolves and dogs (Germonpré et al., 2009, 2012; Germonpré, Lázničkova-Galetová, and Sablin, 2012). The silver fox domestication (or docility breeding) project was started in 1959 by the geneticist Dmitry Belyaev, at the Institute of Cytology and Genetics of the Russian Academy of Sciences (Belyaev, 1969; Trut, 1980, 1999).

### Sexiness and Commitment in the Miocene?

Lovejoy's (2009 [source of the quote on page 62]) model of *Ardipithecus* sociality can be supplemented usefully with discussion of the empirical evidence of the taxon, found in White, Suwa, and Asfaw (1994), WoldeGabriel et al. (1994, 2001, 2009), Haile-Selassie (2001), White et al. (2009a,b), Louchart et al. (2009), Suwa et al. (2009a,b), and Lovejoy et al. (2009a,b,c,d). Germs of the hypothesis appear in Lovejoy (1981) and in Haile-Selassie, Suwa, and White (2004).

Sarmiento (2010; see also Sarmiento, 1987, 1995, 1998, 2000) joins Wood and Harrison (2011) in highlighting the fact that other, non-hominin Miocene apes, such as *Oreopithecus, Ouranopithecus,* and *Sivapithecus (Ramapithecus)*, have small, non-honed canine teeth—supposedly an exclusive hominin trait. In addition, *Oreopithecus,* though not considered a hominin, shows many postcranial skeletal features linked to upright posture and possibly bipedalism, which are often pointed to as exclusively diagnostic of the hominins (Sarmiento, 1987; Harrison and Rook, 1997; Kohler and Moya Sola, 1997). The quote on page 63 is from Begun (2010; see also Begun, 2003, 2007).

### CHAPTER 4: CONCEIVING OUR PAST

A definitive biography of C. K. Brain is yet to be written. A short overview of his life and career was produced by Rubidge (2000), but one gets a better appreciation of Brain's outlook and scientific approach in his own publications, which often contain great anecdotes and convey his love of science and family and friends in clear, elegant prose. I especially recommend *The Hunters or the Hunted? An Introduction to African Cave Taphonomy* (1981a) and *Swartkrans: A Cave's Chronicle of Early Man* (1993). (The information about Attila Port and the tale of Brain's own adventure in a leopard den, as well as the quote on page 68, come from the former book.) Brain, van Riet Lowe, and Dart (1955) described the stone tools that Brain discovered at Makapansgat; the artifacts that he discovered at Sterkfontein were announced in the published version of his doctoral thesis (Brain, 1958). The chapters in Pickering, Schick, and Toth's (2007) edited tribute to Brain demonstrate his continuing influence on taphonomic research in paleoanthropology. Collectively, these three volumes serve as a good introduction to the application of taphonomy in human evolutionary studies.

Lyman's (1994) *Vertebrate Taphonomy* provides encyclopedic coverage of taphonomic topics for the reader interested in exploring advanced applications of the approach in analyzing archaeological bone assemblages; other classic book-length works in paleoanthropological taphonomy include Behrensmeyer and Hill (1980), Shipman (1981), Binford (1981), and Bonnichsen and Sorg (1989). Andrews (1983) set the standard for taphonomic studies of small animal remains collected and modified by owls. Cleghorn and Marean's (2007) useful articulation of a general model of bone portion survivorship in fossil and archaeological assemblages—in which thicker bones of lower nutritional content better survive destructive processes than do ones that are thinner and have higher nutritional value—follows in a long line of successive research on bone density and associated economic value (summarized in Lyman, 1994), which was, in large part, initiated by Bob Brain (1967, 1969, 1981a) early in his taphonomic career (see also especially Bunn, 1986, 1991). Simpson (1970) is a classic paper on actualism and the historical sciences. The original statement of actualism is by Hutton (1794; see also Hutton, 1788), popularized first by Playfair (1802) and best by Lyell (1830–33). Hutton's concept of actualism might fairly be traced, at least in part, to David Hume (1772), who wrote, "For all inferences from experience suppose, as their foundation, that the future will resemble the past, and that similar powers will be conjoined with similar sensible qualities."

Beyond Brain's demolition of the "killer ape hypothesis," Oakley (1956) knocked *Australopithecus africanus/prometheus* down a peg further with his falsification of burning damage on the Makapansgat animal bones (see the chapter 2 text for discussion of Dart's claim that ape-men controlled fire and burned bones at Makapansgat). Chemical analyses did not confirm the presence of free carbon in the bones, and Oakley thus concluded that their blackening was the result of manganese dioxide staining, a common blemish seen on fossils from many South African ape-man caves.

### A Gentler Noble Savagery?

Note that even as early as the 1950s there was minority (but important) opposition to Dart's "killer ape hypothesis." Prominently, Oakley (1954) and Washburn (1957) challenged the notion of australopithecines as masters of their domain. With seminal taphonomic observations that falsified head-hunting by ape-men and pointed instead to their victimization by predators, Washburn was a particularly important influence on Brain's much more fully realized and impactful research; Brain (1981) even appropriated the subtitle of Washburn's 1957 paper, "Australopithecines: The Hunters or the Hunted?," as the title for his own classic book. (Ironically, Washburn [e.g., Washburn and Avis, 1958; Washburn and Lancaster, 1968] was one of the architects of the "man the hunter" paradigm.)

The quote on page 71, demonstrating Dart's magnanimity toward Brain, is from Brain (2001). The quote on page 72 about the "attendant misanthropy" of hunting hypotheses of human origins is from Cartmill (1993). Details on San genetics can be found in: Hammer et al. (2001); Knight et al. (2003); Naidoo et al. (2010). The definitive statement of the "man is hunter" paradigm is Lee and DeVore's (1968) edited volume of the same name. The book is also a good starting point to learn about hunter-gatherer studies and their application to interpreting human evolution. A large literature based on more current research is also available. Useful recent synopses include: Panter-Brick et al. (2001); Lee and Daly (2002); Lamb and Hewlett (2005). Lee (1979, 1982) summarizes statistics and observations of violence among the San groups that he observed in the 1960s. Knauft (1987, 1991) considers interpersonal violence (especially homicide) and sociality in simple human groups. More recent case studies and synthetic considerations of violence and warfare in simple prehistoric and modern societies are found in Keeley (1996), Martin and Frayer (1997), LeBlanc (2003), and Bowles (2009). Abundant cross-cultural data support the notion that human males are, on average, more physically aggressive and murderous than are females; see, for instance, Daly and Wilson (1988) and Björkqvist (1994).

The quote on pages 72–73 is from Haraway (1988). Nancy Tanner and Adrienne Zihlman were among the leading voices in the "woman the gatherer" school (see, e.g., Tanner, 1981; Tanner and Zihlma, 1976). By virtue of its empirical basis, Zihlman et al.'s (1978) bonobo (or pygmy chimpanzee) model of ancestral hominin sociality emerged from the general atmosphere of the 1970s, in which "man the hunter" critiques were ultimately just politicized tracts. However, Zihlman et al.'s bonobo model is also fatally flawed, as revealed by Latimer, White, and Kimbel (1981).

## Back to the Earth

Good technical primers on the identification of butchery marks and of their paleoanthropological interpretations include: Bunn (1981); Shipman and Rose (1983a,b); Blumenschine and Selvaggio (1988); Bonnichsen and Sorg (1989); White (1992); Capaldo and Blumenschine (1994); Pickering and Egeland (2006); Galán et al. (2009); Domínguez-Rodrigo· et al. (2009). Summaries of archaeological evidence for early hominin meat eating include: Domínguez-Rodrigo (2002); Domínguez-Rodrigo and Pickering (2003); Pickering and Domínguez-Rodrigo (2006); Pickering and Bunn (2012). Semaw et al. (2003) and Domínguez-Rodrigo et al. (2005), and de Heinzelin et al. (1999) report, respectively, on the earliest evidence of hominin butchery (and associated stone tools) from the Ethiopian sites of Gona, at 2.6 million years old, and Bouri, at 2.5 million years old. Isaac (1978, 1981, 1984) provides good synopses of his important contributions to Stone Age archaeology, including extended discussion of his "home base" model of early hominin sociality. Bunn's (1981, 1982a, 2001, 2007; Bunn and Kroll, 1986) conclusion about early access to ungulate carcasses by Pleistocene hominins at Olduvai Gorge and Koobi Fora are confirmed and elaborated in book length by Domínguez-Rodrigo, Barba, and Egeland (2007). The most recent summary of archaeological finds and inferences about early hominin behavior at Olduvai Gorge are found in a special issue of the journal *Quaternary Research* (Domínguez-Rodrigo et al., 2010); important earlier works include: L. Leakey (1965); M. Leakey (1971); Hay (1976); Bunn (1982a); Potts (1988).

*Bones: Ancient Men and Modern Myths* is the title of Lewis Binford's (1981 [source of the quote on page 75]) rant against the hypothesis of early hominins as competent hunters. A long-running, often amusing, debate ensued between Binford (1985, 1986, 1988, 1989) and Bunn (1982b, Bunn and Kroll, 1986, 1988). In retrospect, most experts now recognize Bunn's much greater credibility on the taphonomic and archaeological records at Olduvai Gorge and Koobi Fora—records thoroughly investigated by Bunn but not by Binford.

Binford's profound contributions to *archaeological theory* are, however, made clear in several histories of the discipline: see, for example, Willey and Sabloff (1980) and Trigger (2006). Binford and Binford (1968) is an early edited volume introducing core concepts of the New Archaeology and their application to archaeological problems.

A broad consideration of the earliest Stone Age—known as the Oldowan and dating from 2.6–1.5 million years ago—is found in Schick and Toth (1993); a much shorter but also informative summary is by Ambrose (2001). *The Oldowan*, by Toth and Schick (2006), is a recent collection of technical studies. The book also includes an introductory chapter that serves as a comprehensive overview of the temporal and geographic distribution of this first known hominin material culture, as well as a summation of the nuts and bolts of the technology (called "direct hard-hammer percussion"—where, at its most basic application, one stone [called a core] is struck with another [called a hammerstone] to produce the sharp-edged cutting flakes that were then used to butcher carcasses and, presumably, for other tasks as well); see also Toth (1987) and Toth and Schick (2005).

### King of the Beasts: A Munificent Ruler? / Bob and Weave

See Schaller (1972) for the classic study on wild African lion biology and behavior. Turner and Antón (1997) provide an illustrated guide to the evolution, anatomy, and biomechanics of large cats. Ewer (1973) is the definitive overview of African carnivore evolution and contrasts nicely the feeding adaptations of felids and hyenas. Kruuk's (1972) ethological study of spotted hyenas (*Crocuta crocuta*) in East Africa was one of the earliest to show an appreciation of the predatory prowess of hyenas. Mills (1990) compares hunting and other behaviors between Kalahari spotted hyenas and the sympatric brown hyena (*Parahyaena brunnea*). A third extant form of hyena, the striped hyena (*Hyaena hyaena*), is a less avid hunter than are spotted and brown hyenas, and the aardwolf (*Proteles cristata*), a specially derived insectivorous hyena, does not hunt vertebrate prey at all. Contra Dart (1956b [source of the quote on page 77]) and Hughes (1954), the bone-collecting proclivities of hyenas are long well known to biologists and paleoanthropologists alike (see a brief review in Pickering, 2002).

The published output by Rob Blumenschine and his collaborators and students—see, for example, Blumenschine (1986, 1987), Blumenschine, Cavallo, and Caplado (1994), Cavallo and Blumenschine (1989), Capaldo (1997), and Selvaggio (1998)—on passive scavenging by early hominins is prolific and continues even in the face of serious current opposition. Domínguez-Rodrigo's (1999) groundbreaking study on lion feeding was a pivotal contribution in the stepwise falsification of Blumenschine's hypothesis of passive scavenging by hominins. Domínguez-Rodrigo's research dealt with, among other issues, Selvaggio's (1994) propping of the scavenging model when she argued that cut marks resulted not from hominins defleshing intact muscle masses but instead from their removal of marginal scraps of meat left on bones by lions. Cavallo's (1997) own observations of leopard feeding in Tanzania falsified his and

Blumenschine's (Cavallo and Blumenschine, 1989) refinement of the passive scavenging model, in which they argued that hominins pirated carcasses cached in trees by temporarily absent leopards. Capaldo and Peters (1995) provide interesting, but ultimately archaeologically unmatched, data from their taphonomic analysis of wildebeest bone assemblages that are the result of mass drowning events in the Serengeti. The idea of hominins scavenging from sabertooth cat kills was proposed by Marean (1989) and then falsified by his own analysis of a bone assemblage from a sabertooth den (Marean and Erhardt, 1995). Domínguez-Rodrigo and Barba's (2006) evaluation of Blumenschine's tooth mark data through the application of their experimental results on biochemical degradation of bone surfaces is revelatory; it also emphasizes the self-correcting nature of the scientific method in paleoanthropology. For more information on microbial destruction of bone, see Marchiafauai et al. (1974) and Greenlee (1996).

Domínguez-Rodrigo et al. (2012 [the quote on page 83 is from this source]) provide the diagnosis of porotic hyperostosis on the cranial remains of the child from the SHK site at Olduvai Gorge. Because the remains are so fragmentary, it is not possible to assign them to a particular species, but Domínguez-Rodrigo and his colleagues make a circumstantial argument that they are most probably from a *Homo erectus* individual. There is still debate about the exact relationship of porotic hyperostosis and various anemias (El-Najjar, Lozoff, and Ryan, 1975; Palkovich, 1987; Stuart-Macadam, 1987, 1992; Wapler, Crubézy, and Schultz, 2004; Vercellotti et al., 2010), but the condition seems to be produced by the combined effects of hypoferremia and gastrointestinal infections (Walker et al., 2009). Hypoferremia in infants and young children is caused by their ingestion of breast milk that is depleted in vitamin $B_{12}$ or by their lack of access to vitamin $B_{12}$ as they are being weaned. The general importance of iron in the infant diet is discussed in Ryan (1997); see Katzenberg, Herring, and Saunders (1996) for detailed discussion of weaning and its effect on infant mortality from an osteological perspective. Ortner (2003) and Aufderheide and Rodríguez-Martin (2011) are recent primers on skeletal paleopathology. Skeletal indications of anemia, including porotic hyperostosis and cribra orbitalia (a related pathological modification of the eye orbits), increased sharply after the advent of agriculture, which appeared first in Southwest Asia about 11,500 years ago (Price and Bar-Yosef, 2011). These pathologies are rare, but nonetheless still observed on skeletons of modern foragers and prehistoric hunter-gatherers (Lallo, Armelagos, and Mensforth, 1977; Pérez et al., 1997; Bräuer et al., 2003; Vercellotti et al., 2010), people who typically have or had more iron-rich diets than do herders and especially farmers. Blumenschine (1986) and Domínguez-Rodrigo (2001) demonstrate that, in savanna habitats, scavenging opportunities only occasionally yield substantial meat.

### Slinging and Shivving

The world's oldest known wooden spears are from the German Early Paleolithic site of Schöningen, which has also yielded other, rare wooden artifacts, like a presumed throwing stick and handles with slots that once probably held sharp flint blades (Thieme, 1997). The Peninj handaxes are among the oldest known in the world (Isaac and Curtis, 1974), along with those from Konso-Gardula, Ethiopia (Asfaw et al., 1992), and Wonderwerk Cave, South Africa (Chazan et al., 2008), all dating around 1.5 million years old. Recently, even older, nearly 1.8 million-year-old handaxes were announced from the Kenyan site of Kokiselei (Lepre et al., 2011). Like the simple Oldowan tools that predate them, the earliest handaxes were produced by direct hard-hammer percussion (see notes to *Back to the Earth*, a previous section in this chapter), but cores and large flakes were now shaped more specifically by removing flakes on both sides and around the tools' circumferences. Numerous experiments demonstrate that handaxes are efficient for heavy-duty butchery and for woodworking (Jones, 1981; Toth, 1997; Keeley and Toth, 1981), supporting circumstantially the hypothesis that the Peninj handaxes were woodworking tools (the latter reference also includes discussion of the study of the microscopic wear on artifacts from Koobi Fora). The Peninj "handax hypothesis" and phytolith analysis are presented in Domínguez-Rodrigo et al. (2001).

An early, explicit critique of using ethnography in archaeological interpretation is Wobst (1978). Recent counterargument against this negative stance includes Pickering and Bunn (2007) on how the judicious use of ethnographic observations can yield robust and testable hypotheses of early hominin behavior.

The story of Neandertal (the modern spelling of the German word *thal* is *tal*, hence former Neanderthals are now Neandertals in the vernacular, although their Linnaean moniker, *Homo neanderthalensis*, retains the old "th" construction) discovery is well known (Shipman and Trinkaus, 1993; Shreeve, 1996). (My parents would think me remiss if I didn't mention here that the Neander Tal was named for Joachim Neander, a major seventeenth-century hymnist, who wrote, among other church songs, "Praise to the Lord, the Almighty"—and also held religious services in the valley bearing his name.) The recent renaissance in Neandertal genetic studies is unfolding at breakneck speed. For examples of the new research lines and results (including those discussed in this chapter), see Green et al. (2006, 2009, 2010), Noonan et al. (2006), Serre et al. (2006), Wall and Kim (2007), Briggs et al. (2009), and Burbano et al. (2010).

Discussion of the morphologies of Neandertal shoulders and humeri, and their relevance for reconstructing spear use (thrusting versus throwing), is found in Churchill et al. (1996), Schmitt et al. (2003), and Rhodes and Churchill (2009). Trinkaus (2008, 2012; Trinkaus et al., 2006) shows that the shoulders and elbows of Neandertals and later Upper Paleolithic *Homo sapiens* were in some ways significantly alike, observations that call into question whether the two groups had significant differences in their throwing abilities. Churchill (1998, 2006) provides estimates of Neandertal chest depth, and Cowgill (2007) and Larson (2007) discuss Neandertal activity levels as they relate to shoulder and arm morphology. The quote on page 18 is from Schmitt et al. (2003).

For differing influential opinions on the origins of true long-range projectile technology see Brooks et al. (2005) versus Shea (2006). Regardless of the point (or points) of origin for projectile weaponry, archaeologists concur that technology is evident and widespread in the Old World by 50,000–40,000 years ago. Other interesting considerations of Middle Paleolithic (the time period between about 300,000 and 40,000 years ago) stone technology, including that of Neandertals, are Shea (1997) and Shea et al. (2001). More broadly, Knecht (1997) collected informative examples of actualistic and archaeological approaches to studying prehistoric projectile technology. The earliest known atlatls are reported from the French Upper Paleolithic at La Placard (Breuil, 1913) and Combe-Saunière (Cattelin, 1989). Berger and Trinkaus (1995) compared Neandertal skeletal trauma to that of modern rodeo performers; the quote on page 89 is from that source. Trinkaus (2012) emphasizes and elaborates upon his initial caution (Berger and Trinkaus, 1995) in suggesting the "rodeo rider hypothesis." The quote on page 89 is from that source. In addition to the factors discussed in the text, especially relevant to this update are recent analyses of Middle Paleolithic stone artifacts that show morphologies and impact fractures indicating that they were deployed by Neandertals and their contemporaries as projectiles (see, e.g., Boëda et al., 1999; Hardy et al., 2001; Villa et al., 2009; Villa and Soriano, 2010; Lazuén, 2012). Neandertal remains showing injuries that were probably inflicted by other hominins include: Shanidar 3 (Iraq) (Trinkaus, 1983); St. Césaire (France) (Zollikofer et al., 2002); and Sunghir 1 (Russia) (Trinkaus and Buzhilova, 2012).

The Schöningen spears are described by Thieme (1997), the spear from Lehringen by Movius (1950) and that from Clacton-on-Sea by Oakley et al. (1977). Steguweit (1999) conducted experiments showing that the Schöningen spears could have been cast overhand style. To the contrary, Schmitt et al. (2003) critique the idea that they were used as javelins. Gamble (1986, 1987) is responsible for the creative alternative interpretations of the enigmatic wooden artifact fragment from Clacton-on-Sea. See Churchill (1993) and Cattelain (1997) for estimates of effective hunting distances of atlatl- and bow-projected darts and arrows. The quote on page 91 is from Schmitt et al. (2003).

*Man the Ambush Predator*

Tests of the penetrating power of stone-tipped spears and arrows appear in Odell and Cowan (1986) and Huckell (1979, 1982). Stanford (1979) and Frison (1989) threw spears at dead elephants. Manuel Domínguez-Rodrigo kindly shared his unpublished experimental results on the penetration of untipped wooden spears thrown into dead ungulates. The estimates of penetration range summarized in the text are from him.

Ruff et al. (1997) estimate that typical individuals of nonmodern, Pleistocene *Homo* species were heavily muscled. We also assume that since these early humans were subsistence foragers, they would have, for the most part, been in good physical shape.

Larson (2007, 2009) provides detailed comparisons of shoulder morphology in the genus *Homo*, including discussion of the anatomy of Nariokotome Boy's shoulder girdle and the similarities of his collarbones to those of modern humans with short clavicle syndrome. (Perhaps relevant here, as discussed in chapter 1, at least one assessment of Nariokotome Boy's thoracic skeleton concludes that the boy suffered from pathological axial dysplasia

[Latimer and Ohman, 2001].) The new Dmanisi *Homo erectus* collarbones, one from a juvenile skeleton and two (right and left) from a single adult skeleton, are described in Lordkipanidze et al. (2007).

Zoo keepers (and even some unlucky patrons) may attest to the throwing proclivities of captive chimpanzees—often with creatively repulsive choices of projectiles and disappointingly good aim! There are also reports of wild chimpanzees throwing sticks and rocks in defensive actions against predators, like leopards and lions (Nishida, 1968; van Lawick-Goodall, 1968; Hiraiwa-Hasegawa et al., 1986), and at food competitors, like baboons (Goodall, 1964). Kanzi, the well-known captive bonobo, makes sharp-edged stone cutting flakes by throwing rocks against the concrete floor of his enclosure (Toth et al., 1993; Schick et al., 1999).

Evidence of hominin-controlled fire predating 790,000 years old (Goren-Inbar et al., 2004) is arguably equivocal and always contentious. Common weaknesses include the presence of questionably burned materials in good archaeological contexts; the presence of unquestionably burned materials (sediments, bones, seeds) in questionable archaeological and/or temporal contexts; or, the presence of unquestionably burned materials in good archaeological contexts but that were not definitively burned by intentional hominin activity. Klein (1999; see also Roebroeks and Villa, 2011) summarizes critiques of the contenders for the earliest domestication of fire at sites like FxJj 20 East (Koobi Fora, Kenya) (Bellomo, 1994), Chesowanja (Kenya) (Gowlett et al., 1981), Gadeb 8e (Middle Awash, Ethiopia) (Clark and Harris, 1985), and Swartkrans Cave (South Africa) (Brain and Sillen, 1988). Berna et al. (2012) describe and discuss the evidence of controlled, in situ fires in the one million-year-old levels of Wonderwork Cave (South Africa). Some commentators are even wary of the Wonderwerk claims, their nonreviewed, online opinions stressing the lack of defined hearths in the cave. The earliest verified, intentionally heat-treated artifacts are stone projectile points from Pinnacle Point, South Africa, recovered from a relatively recent archaeological level, dating to about 164,000 years ago (Brown et al., 2009).

Churchill (1993) provides important comparative data on the hunting techniques of modern foragers. See also Ellis (1997) for useful comparative data on traditional projectile hunting; the quote on page 99 is from this source.

The "endurance running hunting hypothesis" (Bramble and Lieberman, 2004) garnered wide and enthusiastic press (including its promotion in Christopher McDougall's [2009] recent popular book, *Born to Run*), but severe criticism of the hypothesis, based on the archaeological and paleoecological records of early *Homo* (Pickering and Bunn, 2007), as well as on that hominin's functional anatomy (Simpson et al., 2008) and reconstructed physiology (Steudel-Numbers and Wall-Scheffler, 2009), followed; the quote on page 101 is from the last source. Pickering and Bunn's critiques are discussed in the text of chapter 4; see also Cerling, Bowman, and O'Neil (1988), Cerling (1992), Sikes (1994), Reed (1997), and Cerling et al. (2011) for examples of studies that reconstruct the habitat of early *Homo* as predominantly savanna-woodland. Simpson and his colleagues demonstrated that the new female presumptive *Homo erectus* pelvis from Busidima, Ethiopia, shows none of the derived morphology seen in modern human hips, which are much narrower side to side than are the hips of more archaic hominin forms. Based on this interpretation, the idea that the narrow human pelvis originated in *Homo erectus* as an adaptation to endurance running is falsified. But, as discussed in the notes to chapter 1, see Ruff's (2010) argument that the Busidima pelvis might be that of *Australopithecus* rather than *Homo erectus*. If Ruff's taxonomic assessment is correct, then Busidima is obviously of no direct relevance to reconstructing the hips of *Homo erectus*. Steudel-Numbers and Wall-Scheffler studied the energetics of modern human running in order to evaluate the "endurance running hypothesis." They showed that hunting by running, even at optimal speed, would have been extremely energetically costly to a hominin predator chasing typical ungulate prey. Liebenberg (2006) reports on the twenty San persistence hunts.

See Bunn (1982a, 1986, 1991, 2007; Bunn and Kroll, 1986) and Domínguez-Rodrigo et al. (2007) for the paleoanthropological relevance of FLK Zinj, with special emphasis on the compelling zooarchaeological and taphonomic evidence that the site's hominins butchered choice portions of ungulate prey carcasses. (Based on initial results, we predict that the site of BK, also at Olduvai but slightly younger than FLK Zinj, will soon overtake FLK Zinj as the site with the most and best evidence of early hominin meat eating and hunting [see, e.g., Monahan, 1996; Egeland, 2007; Egeland and Domínguez-Rodrigo, 2008; Domínguez-Rodrigo et al., 2009a; Pickering et al., 2013].) Bunn and Pickering (2010a,b) present the ungulate mortality data for FLK Zinj, which leads to a hypothesis of ambush hunting by early *Homo;* the quote on page 102 is from Pickering and Bunn (2012). Analyzing the mortality distribution of prey animals is a well-established archaeological method to assess hominin hunting competence (and, in more recent archaeological contexts of early agriculture, herd management strategies): see general discussions of mortality analysis in Lyman (1994), and seminal applications to paleoanthropology by Klein (1978, 1999). Stiner (1990), Steele and Weaver (2002), and Bunn and Pickering (2010a,b) update these earlier studies methodologically.

For discussions of African ape philopatry and its evolutionary context and implications, see: Itani (1985); Wrangham (1986); Pusey and Packer (1987); Furuichi (2006). In contrast to the dispersal of females from natal groups that is observed in chimpanzees and bonobos, data reveal that both sexes disperse in some gorilla populations (Robbins et al., 2004). Lovejoy (1981, 2009d), Ghiglieri (1987), and Wrangham (1987) reconstruct the socioecology of the most recent common ancestor of African apes and humans and also discuss implications of that ancestor's inferred philopatry. Boehm (1999) also argues effectively that, for traditional band and tribal people, within-group coalitions are critical to holding dominant individuals in check. Coalitions of foraging male primates might also function to provide safety from predators.

## CHAPTER 5: DEATH FROM ABOVE

In addition to proffering opinion about its evolutionary position, Asfaw et al. (1999) announced and described the anatomy of *Australopithecus garhi*. De Heinzelin et al. (1999) presented the evidence of butchery by early hominins at Bouri, and Semaw et al. (2003) and Domínguez-Rodrigo et al. (2005) described the evidence from Gona. See Bunn (2007) versus Pickering and Egeland (2009; see also Pickering and Bunn, 2012) for contrasting opinions about the behavioral significance of the Bouri and Gona butchery evidence. Bunn is dubious that this earliest data of hominin interaction with ungulate carcasses also constitutes *compelling* evidence for well-established, humanlike hunting abilities during this phase of prehistory; Pickering and Egeland are more open to a generous assessment of the limited data.

Although it is ephemeral compared to that of humans, chimpanzees possess a rich and diverse material culture (reviewed in McGrew, 1992, 2004; Whiten et al., 1999, 2001; Whiten, Horner, and Marshall-Pescinin, 2003; Whiten, 2005; Whiten, Horner, and de Waal, 2005). They might also produce archaeological traces of behavior that are perhaps detectable over at least thousands of years (Mercader et al., 2007). Chimpanzee wooden digging sticks and stone pounding tools are, in particular, relatively durable and varied in form and use across populations; for examples, see Jones and Sabater Pí (1969), McGrew et al. (1979), Boesch and Boesch (1984, 1990), Goodall (1986), Bermejo and Illera (1999), Boesch and Boesch-Achermann (2000), Mercader, Panger, and Boesch (2002), Sanz et al. (2004), Heaton and Pickering (2006), and Whiten, Schick, and Toth (2009).

The debate (thus far) over the veracity of purported butchery marks on 3.4 million-year-old fossils from Dikika, Ethiopia, is found (chronologically) in: McPherron et al. (2010); Domínguez-Rodrigo, Pickering, and Bunn (2010); McPherron et al. (2011); Domínguez-Rodrigo, Pickering, and Bunn (2011, 2012). (Several of our colleagues, including, prominently, Tim White and Richard Klein, have lodged informal doubts about the authenticity of the Dikika "butchery marks" in the secondary and tertiary literature.) Domínguez-Rodrigo et al. (2009b) is a comprehensive experimental study comparing stone tool butchery marks to marks imparted on bone surfaces incidentally as they were ground against and within encasing sediments upon which they were deposited. The results are illuminating and particularly relevant to our misgivings about the remarkable claims for early butchery from Dikika. Other important studies of processes that can create bone surface damage that mimics butchery marks, include: Shipman and Rose (1984); Andrews and Cook (1985); Behrensmeyer et al. (1986, 1989); Olsen and Shipman (1988); Fiorillo (1989); Oliver (1989).

### Greco-Roman Ape-Men

Newton-Fisher (2007) reviews hypotheses of why chimpanzees hunt, which include models in which hunting serves an intrasexual social function, with participating males creating alliances centered around hunting and meat sharing (Mitani and Watts, 2001). In contrast, proponents of the idea that chimpanzees hunt, ultimately, in order to supplement protein and fat requirements when these microresources are unavailable seasonally in other types of foods include: Takahata et al. (1984); Stanford et al. (1994); Pickering and Domínguez-Rodrigo (2010, 2012). Domínguez-Rodrigo's (2001; see also Blumenschine,

1986) field observations in East Africa demonstrate the ephemeral nature of scavengeable carcass parts in savanna habitats. Brain (2007 [source of the quote on page 111]) tells the touching and amusing story of confronting Raymond Dart with evidence of the ordinary, but actual bone tool tradition that he, Brain, recognized at Swartkrans.

The quote on page 112 is from Goodall (1986). Carrier's (2007) argument, that the short legs of ape-men indicate their high levels of aggression, is published in the journal *Evolution*. White et al.'s (2009a) intimation that large male body size in more recent species of *Australopithecus* might have evolved in response to predation pressure is accompanied by further supposition that increased female cooperation and group cohesion in those species may have, in concert, reduced female body size. If true, these two factors combined could explain elevated levels of body size sexual dimorphism in derived ape-men species without having to appeal to the standard explanation of intragroup male–male aggression (for the detailed consideration of that standard explanation, see: Clutton-Brock, 1985; Leigh, 1997; McHenry, 1994; Plavcan and van Schaik, 1997; Plavcan, 2000, 2001). A recent study of the strontium isotopes of *Australopithecus africanus* and *Australopithecus robustus*, derived hominin species from South Africa, seemed to indicate that males of these species did *not* forage widely on the landscape but instead confined their activities in proximity of the caves in which their fossils were recovered (Copeland et al., 2011). However, a major analytical foundation of this study was proved to be seriously flawed (Plavcan, 2012).

*Unholy Eucharist?*

Dugard (1995; see also Clarke, 1998, 1999, 2002, 2007, 2008) recounts the early history of "Little Foot's" discovery at Sterkfontein Cave, but the story continues to unfold even today. Little Foot joins Ardi (White et al., 2009a), Lucy (Johanson et al., 1982), two partial skeletons from Sterkfontein (catalog numbers Sts 14 [Robinson, 1972] and StW 431 [Toussaint et al., 2003]), the remains of a three-year-old *Australopithecus afarensis* girl from Dikika, Ethiopia (Alemseged et al., 2006), two (maybe four) partial skeletons designated as *Australopithecus sediba*, from the South African site of Malapa, (Berger et al., 2010), and a partial *Australopithecus afarensis* skeleton from Woranso-Mille, Ethiopia (Haile-Selassie et al., 2010), as the most complete pre-*Homo* hominin remains from the African fossil record.

In order to distinguish it from the better-represented and co-occurring species, *Australopithecus africanus* (the Taung Child's species, first discovered and named by Raymond Dart), Clarke (1988, 1994b, 2008) previously referred to the taxon that he argues is represented by Little Foot and by several other fossils from Sterkfontein and Makapansgat Cave as the "second species." More recently, Clarke (e.g., 2013) suggests that second species be given the taxonomic designation *Australopithecus prometheus*, the original name bestowed by Raymond Dart to the Makapansgat ape-man fossils. I also employ the name *Australopithecus prometheus* for my discussions of the "second species" here and in the text. Unlike *Australopithecus africanus*, *Australopithecus prometheus*—with its forward-shifted cheek bones and its enormous premolars and molars that have bulbous cusps—seems to anticipate the anatomy of the highly specialized robust australopithecines (see text of chapter 2). Clarke (1988) is so impressed by these morphological continuities that he argues *Australopithecus prometheus* is "ideally placed morphologically and temporally to be a member of the species ancestral to and directly on the lineage of [the robust ape-men]."

Hughes and Tobias (1977) announced and described the StW 53 skull and were the first to assign it to the genus *Homo;* subsequent analyses agreed with that attribution (Howell, 1978). Clarke's (2008; Kuman and Clarke, 2000) objection, placing the skull instead in the genus *Australopithecus,* is well published. Indeed, general uncertainty reigns over the taxonomy of presumptive early *Homo* and presumptive transitional-to-*Homo* australopithecine species. Relevant discussions concerning the latter group of species, which includes *Australopithecus garhi* and *Australopithecus sediba,* are sketched in the chapter 2 notes. For a broader historical perspective, I summarize more of the debate over early *Homo* here.

*Homo habilis* was found at Olduvai Gorge in 1964 and recognized immediately as contemporary with but—because of its inferred large brain and smaller cheek teeth—different than *Australopithecus boisei* (Leakey et al., 1964). The archaeological association of *Homo habilis* remains with stone tools led Louis Leakey and other early researchers to relegate *Australopithecus boisei* (which Leakey first hypothesized to be a direct ancestor of modern humans) to a side branch in the human family tree and to elevate *Homo habilis,* in its place, to the position of a direct ancestor of *Homo sapiens.* As more presumptive *Homo habilis* fossils were discovered over the years, some researchers were impressed by the wide range of morphological variation evinced in the expanded sample. In particular, Bernard Wood (e.g., 1991, 1992, 1993) divided the collection into two distinct species, one with a smaller brain and smaller teeth who retained the name *Homo habilis,* and one with a larger brain and larger teeth, *Homo rudolfensis.* More recently, Wood (e.g., Wood and Collard, 1999; Collard and Wood, 2007; Wood and Leakey, 2011) has gone further, arguing that *Homo habilis* and *Homo rudolfensis* both share so many morphological adaptations with ape-men that it might be justifiable to reassign them to the genus *Australopithecus.*

Back to Sterkfontein and the gracile australopithecine StW 53: the stratigraphic reinterpretation of StW 53 is found in Kuman and Clarke (2000) and in Clarke (1994c). Probable stone tool cut marks on the facial skeleton of St W 53 (Pickering et al., 2000), if the result of cannibalism, stand as the world's oldest known example of anthropophagy, dating to perhaps 2 million years ago. It is not, however, the only example of probable prehistoric cannibalism. In addition to butchery damage on the Bodo cranium (White, 1986), and on the remains of early modern humans from Ethiopia (Clark et al., 2003) (see chapter text for discussion), the pattern of hominin-on-hominin butchery is confirmed as a relatively common occurrence in nearly all stages of human prehistory, and in almost worldwide distribution from the Pleistocene onward (see, e.g., Villa et al., 1986; White, 1992; Defleur et al., 1993, 1999; Fernández-Jalvo et al., 1996, 1999; Turner and Turner, 1998; Ramirez Rossi et al., 2009; Scott and

Marean, 2009). The full title of Cormac McCarthy's (1985) masterpiece is *Blood Meridian, or the Evening Redness in the West; Blood Meridian* is the source of many wished for epigraphs for this book but permission was denied.

See Clarke (2013) for his opinion that the marks on StW 53 are of natural origin; the quote on page 117 is from that source. The quote on page 118 is from an e-mail Clarke sent to me on October 20, 2011. The quote on page 118 is from Pickering (1999).

## Blitz

Ancient disease in early hominins is detectable, or at least suggested, in some cases. One of the best-known approaches for detecting disease from skeletal remains is the study of hypoplasic defects on teeth. Hypoplasias are disruptions in the surface contour of a tooth's crown, which are expressed in different ways, including as pits and furrows or as simple discoloration (Hillson, 1986). Hypoplasias are distinct and distinguishable from a tooth's normal growth bands, the perikymata (see text of chapter 1), and have been recognized on the teeth of various early hominin species. For examples of paleoanthropologically focused studies of hypoplasias, see: Robinson (1956); White (1978); Molnar and Molnar (1985); Tobias (1991); Bermédez De Castro and Pérez (2003); Guatelli-Steinberg (2003). Hypoplasias are accurate indicators of physical stress at points in an individual's life—but, unfortunately, it is impossible to link the occurrence of a hypoplasia to the specific source of stress that an extinct hominin experienced. In modern humans, it *is* known that these enamel defects are associated with malnutrition and various infectious diseases (Skinner and Goodman, 1992). Rarely, dental caries (or cavities) are seen on the teeth of ape-men (see, though, e.g., Robinson, 1952; Tobias, 1974; Grine et al., 1990); a single juvenile specimen of *Australopthecus africanus* has modification of its bony jaw tissue indicating generalized periodontitis (Ripamonti et al., 1997); and septicemia from a tooth abscess (apparent as a lesion on the right side of the skeleton's lower jawbone) likely caused the demise of the *Homo erectus* Nariokotome Boy (Walker, 1993). Another (partial) skeleton of *Homo erectus*, cataloged as KNM ER 1808, displays abnormal apposition of bone on the shafts of its arm and leg fossils. The etiology of such pathology among modern humans is associated with a disease called yaws, which is related to syphilis, or with hypervitaminosis A. Both causes have been suggested to explain the unusual growths on the bones of KNM ER 1808 (Rothschild et al., 1995; Walker, Zimmerman, and Leakey, 1982). How KNM ER 1808 may have developed hypervitaminosis A—if indeed it did—is further debated. A poisonous buildup of vitamin A is known to occur after frequent consumption of animal (especially carnivore) livers (Walker, Zimmerman, and Leakey, 1982). Alternatively, highly dangerous concentrations of vitamin A occur in the broods of honeybees, which may have been targeted as food by *Homo erectus*, as they are by modern hunter-gatherers (Skinner, 1991). The hypothesis of brucellosis in *Australopithecus africanus* (D'Anastasio et al., 2009) is discussed in the text of chapter 2. The hypothesis of anemia in a hominin child from Olduvai Gorge, Tanzania (Domínguez-Rodrigo et al., 2012), is discussed in the text of chapter 4. Evidence of carnivore tooth marks on fossils of *Australopithecus robustus, afarensis, anamensis,* and *africanus* and on those of *Orrorin* and *Ardipithecus* is presented, respectively, in: Brain (1970, 1974); Johanson et al. (1982); Ward et al. (1999); Pickering, Clarke, and Moggi-Cecchi (2003); Pickford and Senut (2001); WoldeGabriel et al. (1994).

Berger and Clarke (1995) present the Taung Child "raptor hypothesis" and summarize the nature of the fossil assemblage recovered in context with the hominin baby, as well as previous speculation about its depositional history (see also Dart, 1926, 1929; Broom, 1934; Broom and Schepers, 1946). Berger and Clarke (1996) dispelled early reaction that an eagle would have been incapable of lifting the child (Hedenstrom, 1996). Subsequently, Berger (2006; Berger and McGraw, 2007) bolstered the raptor hypothesis with more detailed study of bone surface damage on the child's skull—damage that matches that incurred on the skulls of monkeys that were killed and fed on by modern eagles. The "raptor hypothesis" spawned great interest in the topic as well as a spate of actualistic work on the taphonomy of African raptors, especially with regard to the accumulation of primate remains (see Sanders et al., 2003; McGraw et al., 2006; Trapani et al., 2006; Gilbert et al., 2009). The quote on page 123 is from Dart (1953).

**CODA**

James Ellroy's (1996) quote on page 125 is from his autobiography, *My Dark Places*. The translated quote on pages 126–127 is from de Maistre (1965; see de Maistre 1836 for original French version). Chris Hedge's (2003) views on war are well expressed in *War Is a Force That Gives Us Meaning*.

Samuel Bowles's contributions to understanding the evolutionary bases of human morality are many and important. With regard to the discussion of how human warfare evolved, see, in particular, Bowles (2009) and Bowles and Gintis (2011). The quote on pages 126–127 is from Richerson (2011).

# References

Aiello, L. C. 2007. "Notes on the Implications of the Expensive Tissue Hypothesis for Human Biological and Social Evolution." In *Guts and Brains: An Integrated Approach to the Hominin Record*, edited by W. Roebroeks, 17–28. Leiden, The Netherlands: Leiden University Press.

Aiello, L. C., Bates, N., and Joffe, T. 2001. "In Defense of the Expensive Tissue Hypothesis." In *Evolutionary Anatomy of the Primate Cerebral Cortex*, edited by D. Falk and K. Gibson, 57–78. Cambridge, UK: Cambridge University Press.

Aiello, L. C., and Wells, J. C. K. 2002. "Energetics and the Evolution of the Genus *Homo.*" *Annual Review of Anthropology* 31: 323–38.

Aiello, L. C., and Wheeler, P. 1995. "The Expensive Tissue Hypothesis: The Brain and Digestive System in Human and Primate Evolution." *Current Anthropology* 36: 199–221.

Alemseged, Z. et al. 2006. "A Juvenile Early Hominin Skeleton from Dikika, Ethiopia." *Nature* 443: 296–301.

Alexander, R. D. 1974. "The Evolution of Social Behavior." *Annual Review of Ecology and Systematics* 5: 325–84.

Alexander, R. D. 1987. *The Biology of Moral Systems*. New York: Aldine de Gruyter.

Ambrose, S. H. 2001. "Paleolithic Technology and Human Evolution." *Science* 291: 1748–53.

Anderson, D. L., Thompson, G. W., and Popovich, F. 1976. "Age Attainment of Mineralization Stages of the Dentition." *Journal of Forensic Science* 21: 191–200.

Andrews, P. 1983. *Owls, Caves and Fossils*. Chicago: University of Chicago Press.

Andrews, P., and Cook, J. 1985. "Natural Modifications to Bones in a Temperate Setting." *Man* 20: 675–91.

Antón, S. C. 2003. "Natural History of *Homo erectus.*" *Yearbook of Physical Anthropology* 46: 126–69.

Ardrey, R. 1961. *African Genesis.* New York: Atheneum.

Ardrey, R. 1970. *The Social Contract.* New York: Atheneum.

Ardrey, R. 1971. *The Territorial Imperative.* New York: Dell.

Ardrey, R. 1976. *The Hunting Hypothesis.* New York: Macmillan.

Asfaw, B. et al. 1992. "The Earliest Acheulean from Konso-Gardula." *Nature* 360: 732–34.

Asfaw, B. et al. 1999. "*Australopithecus garhi:* A New Species of Early Hominid from Ethiopia." *Science* 284: 629–34.

Aufdeheide, A. C., and Rodríguez-Martin, C. 2011. *The Cambridge Encyclopedia of Human Paleopathology.* Cambridge, UK: Cambridge University Press.

Backwell, L. R., and d'Errico, F. 2001. "First Evidence of Termite-Foraging by Swartkrans Early Hominids." *Proceedings of the National Academy of Science (USA)* 98: 1358–63.

Bardeleben, C., Moore, R. L., and Wayne, R. K. 2005. "Isolation and Molecular Evolution of the Selenocysteine tRNA (*CF TRSP*) and RNase P RNA (*CF RPPH1*) Genes in the Dog Family, Canidae." *Molecular Biology and Evolution* 22: 347–59.

Barkow, J. H., Cosmides, L., and Tooby, J. (editors). 1992. *The Adapted Mind: Evolutionary Psychology and the Generation of Culture.* New York: Oxford University Press.

Becquet, C., and Przeworski, M. 2007. "A New Approach to Estimate Parameters of Speciation Models with Applications to Apes." *Genome Research* 17: 1505–19.

Begun, D. R. 2003. "Planet of the Apes." *Scientific American* 289: 74–83.

Begun, D. R. 2007. "Fossil Record of Miocene Hominoids." In *Handbook of Palaeoanthropology, Vol. 2: Primate Evolution and Human Origins,* edited by W. Henke and I. Tattersall, 921–77. Berlin: Springer.

Begun, D. R. 2010. "Miocene Hominids and the Origins of the African Apes and Humans." *Annual Review of Anthropology* 39: 67–84.

Behrensmeyer, A. K., Gordon, K. D., and Yanagi, G. T. 1986. "Trampling as a Cause of Bone Surface Damage and Pseudo-Cutmarks." *Nature* 319: 768–71.

Behrensmeyer, A. K., Gordon, K. D., and Yanagi, G. T. 1989. "Nonhuman Bone Modification in Miocene Fossils from Pakistan." In *Bone Modification,* edited by R. Bonnichsen and M. Sorg, 99–120. Orono, ME: Center for the Study of the First Americans.

Behrensmeyer, A. K., and Hill, A. P. (editors). 1980. *Fossils in the Making: Vertebrate Taphonomy and Paleoecology.* Chicago: University of Chicago Press.

Bellomo, R. V. 1994. "Methods of Determining Early Hominid Behavioral Activities Associated with the Controlled Use of Fire at FxJj 20 Main, Koobi Fora, Kenya." *Journal of Human Evolution* 27: 173–95.

Belyaev, D. K. 1969. "Domestication of Animals." *Science Journal* 5: 47–52.

Berger, L. R. 2006. "Predatory Bird Damage to the Taung Type-Skull of *Australopithecus africanus* Dart 1925." *American Journal of Physical Anthropology* 13: 166–68.

Berger, L. R., and Clarke, R. J. 1995. "Eagle Involvement in Accumulation of the Taung Child Fauna." *Journal of Human Evolution* 29: 275–99.

Berger, L. R., and Clarke, R. J. 1996. "The Load of the Taung Child." *Nature* 379: 778–79.

Berger, L. R., and McGraw, W. S. 2007. "Further Evidence for Eagle Predation of, and Feeding Damage on, the Taung Child." *South African Journal of Science* 103: 496–98.

Berger, L. R. et al. 2010. "*Australopithecus sediba*: A New Species of *Homo*-like Australopith from South Africa." *Science* 328: 195–204.

Berger, T. D., and Trinkaus, E. 1995. "Patterns of Trauma among the Neandertals." *Journal of Archaeological Science* 6: 101–4.

Bermédez De Castro, J. M., and Pérez, P.-J. 2003. "Enamel Hypoplasia in the Middle Pleistocene Hominids from Atapuerca (Spain)." *American Journal of Physical Anthropology* 96: 301–14.

Bermejo, M., and Illera, G. 1999. "Tool-Set for Termite Fishing and Honey Extraction by Wild Chimpanzees in the Lossi Forest, Congo." *Primates* 40: 619–27.

Berna, F. et al. 2012. "Microstratigraphic Evidence of *in situ* Fire in the Acheulean Strata of Wonderwerk Cave, Northern Cape Province, South Africa." *Proceedings of the National Academy of Sciences (USA)* 109: E1215–20.

Beynon, A. D., and Dean, M. C. 1988. "Distinct Dental Development Patterns in Early Fossil Hominids." *Nature* 335: 509–14.

Beynon, A. D., and Wood, B. A. 1987. "Patterns and Rates of Enamel Growth on the Molar Teeth of Early Hominids." *Nature* 326: 493–96.

Binford, L. R. 1981. *Bones: Ancient Men and Modern Myths*. New York: Academic Press.

Binford, L. R. 1985. "Human Ancestors: Changing Views of their Behavior." *Journal of Anthropological Archaeology* 4: 292–327.

Binford, L. R. 1986. "Comment on Bunn and Kroll." *Current Anthropology* 27: 444–46.

Binford, L. R. 1988. "Fact and Fiction about the *Zinjanthropus* Floor: Data, Arguments and Interpretations." *Current Anthropology* 29: 123–35.

Binford, L. R. 1989. *Debating Archaeology*. San Diego: Academic Press.

Binford, S. R., and Binford, L. R. (editors). 1968. *New Perspectives in Archaeology*. Chicago: Aldine.

Björkqvist, K. 1994. "Sex Differences in Physical, Verbal, and Indirect Aggression: A Review of Recent Research." *Sex Roles* 30: 177–88.

Blumenschine, R. J. 1986. "Early Hominid Scavenging Opportunities: Implications of Carcass Availability in the Serengeti and Ngorongoro Ecosystems." *British Archaeological Reports International Series* 283. Oxford, UK: Archaeopress.

Blumenschine, R. J. 1987. "Characteristics of an Early Hominid Scavenging Niche." *Current Anthropology* 28: 383–407.

Blumenschine, R. J., Cavallo, J. A., and Caplado, S. D. 1994. "Competition for Carcasses and Early Hominid Behavioral Ecology: A Case Study and Conceptual Framework." *Journal of Human Evolution* 27: 197–213.

Blumenschine, R. J., and Selvaggio, M. 1988. "Percussion Marks on Bone Surfaces as a New Diagnostic of Hominid Behavior." *Nature* 333: 763–65.

Blumenschine, R. J. et al. 2003. "Late Pliocene *Homo* and Land Use from Western Olduvai Gorge, Tanzania." *Science* 299: 1217–21.

Bobe, R., Zeresenay, A., and Behrensmeyer, A. K. (editors). 2007. *Hominin Environments in the East African Pliocene: An Assessment of the Faunal Evidence*. Dordrecht, The Netherlands: Springer.

Boëda, E. et al. 1999. "A Levallois Point Embedded in the Vertebra of a Wild Ass (*Equus africanus*): Hafting, Projectiles and Mousterian Hunting Weapons." *Antiquity* 3: 394–402.

Boehm, C. 1999. *Hierarchy in the Forest: The Evolution of Egalitarian Behavior*. Cambridge, MA: Harvard University Press.

Boesch, C., and Boesch, H. 1984. "Mental Map in Wild Chimpanzees: An Analysis of Hammer Transport for Nut-Cracking." *Primates* 25: 160–70.

Boesch, C., and Boesch, H. 1989. "Hunting Behavior of Wild Chimpanzees in the Taï National Park." *American Journal of Physical Anthropology* 78: 547–73.

Boesch, C., and Boesch, H. 1990. "Tool Use and Tool Making in Wild Chimpanzees." *Folia Primatologia* 54: 86–99.

Boesch, C., and Boesch-Achermann, H. 2000. *The Chimpanzees of the Taï Forest: Behavioural Ecology and Evolution*. Oxford, UK: Oxford University Press.

Boesch, C., Hohmann, G., and Marchant, L. (editors). 2001. *Behavioral Diversity in Chimpanzees and Bonobos*. Cambridge, UK: Cambridge University Press.

Bogin, B. 1999. *Patterns of Human Growth*, 2nd ed. Cambridge, UK: Cambridge University Press.

Bogin, B., and Smith, B. H. 2000. "Evolution of the Human Life Cycle." In *Human Biology: An Evolutionary and Biocultural Perspective*, edited by S. Stinson et al., 377–424. New York: Wiley-Liss.

Bonnichsen, R., and Sorg, M. H. (editors). 1989. *Bone Modification*. Orono, ME: Center for the Study of the First Americans.

Bowles, S. 2009. "Did Warfare among Ancestral Hunter-Gatherers Affect the Evolution of Human Social Behaviors?" *Science* 324: 1293–98.

Bowles, S., and Gintis, H. 2011. *A Cooperative Species: Human Reciprocity and Its Evolution*. Princeton, NJ: Princeton University Press.

Brace, C. L. 1967. *The Stages of Human Evolution*. Englewood Cliffs, NJ: Prentice Hall.

Bradley, B. 2008. "Reconstructing Phylogenies and Phenotypes: A Molecular View of Human Evolution." *Journal of Anatomy* 212: 337–53.

Brain, C. K. 1958. *The Transvaal Ape-Man-Bearing Cave Deposits*. Pretoria: Transvaal Museum.

Brain, C. K. 1967. "Hottentot Food Remains and Their Bearing on the Interpretation of Fossil Bone Assemblages." *Scientific Papers of the Namib Desert Research Station* 29: 13–22.

Brain, C. K. 1969. "The Contributions of Namib Desert Hottentots to an Understanding of Australopithecine Bone Accumulations." *Scientific Papers of the Namib Desert Research Station* 32: 1–11.

Brain, C. K. 1970. "New Finds at the Swartkrans Australopithecine Site." *Nature* 225: 1112–19.

Brain, C. K. 1974. "A Hominid Skull's Revealing Holes." *Natural History* 83: 44–45.

Brain, C. K. 1981a. *The Hunters or the Hunted? An Introduction to African Cave Taphonomy.* Chicago: University of Chicago Press.

Brain, C. K. 1981b. "The Evolution of Man in Africa: Was It a Consequence of Cenozoic Cooling?" Alex L. Du Troit Memorial Lecture no. 17. *The Geological Society of South Africa, Annexure to Vol. 84:* 1–19.

Brain, C. K. 1984. "The Terminal Miocene Event: A Critical Environmental and Evolutionary Episode?" In *Late Cenozoic Palaeoclimates of the Southern Hemisphere,* edited by J. C. Vogel, 491–98. Rotterdam and Boston: A. A. Balkema.

Brain, C. K. (editor). 1993. *Swartkrans: A Cave's Chronicle of Early Man.* Pretoria: Transvaal Museum.

Brain, C. K. 2001. "Do We Owe Our Intelligence to a Predatory Past?" Seventieth James Arthur Lecture on the Evolution of the Human Brain. New York: American Museum of Natural History.

Brain, C. K. 2007. "Fifty Years of Fun with Fossils: Some Cave Taphonomy-Related Ideas and Concepts That Emerged between 1953 and 2003." In *Breathing Life into Fossils: Taphonomic Studies in Honor of C. K. (Bob) Brain,* edited by T. R. Pickering, K. Schick, and N. Toth, 1–24. Bloomington, IN: Stone Age Institute Press.

Brain, C. K., and Shipman, P. 1993. "The Swartkrans Bone Tools." In *Swartkrans: A Cave's Chronicle of Early Man,* edited by C. K. Brain, 195–215. Pretoria: Transvaal Museum.

Brain, C. K., and Sillen, A. 1988. "Evidence from Swartkrans Cave for the Earliest Use of Fire." *Nature* 336: 464–96.

Brain, C. K., van Riet Lowe, C., and Dart, R. A. 1955. "Kafuan Stone Artefacts in the Post-Australopithecine Breccia at Makapansgat." *Nature* 175: 16–18.

Brain, C. K. et al. 1988. "New Evidence of Early Hominids, Their Culture and Environment from the Swartkrans Cave, South Africa." *South African Journal of Science* 84: 828–35.

Brain, C. K. et al. 2001. "Interpretive Problems in a Search for Micro-invertebrate Fossils from a Neoproterozoic Limestone in Namibia." *Palaeontologia Africana* 37: 1–12.

Brain, C. K. et al. 2003. "Sponge-Like Microfossils from Neoproterozoic Intertillite Limestones of the Otavi Group in Northern Namibia." In *International Geological Correlation Programme, Project 478: Neoproterozoic-early Paleozoic Events in SW-Gondwana,* edited by H. E. Frimmel, 19–22. Cape Town: IGCP.

Brain, C. K. et al. 2012. "The First Animals: Ca. 760-Million-Year-Old Sponge-Like Fossils from Namibia." *South African Journal of Science* 108: 1–8.

Bramble, D. M., and Lieberman, D. E. 2004. "Endurance Running and the Evolution of *Homo.*" *Nature* 432: 345–52.

Bräuer, G. et al. 2003. "Pathological Alterations in the Archaic *Homo sapiens* Cranium from Eliye Springs, Kenya." *American Journal of Physical Anthropology* 120: 200–204.

Breuil, H. 1913. "Les subdivisions du Paléolithique supérieur et leur significa-tion." XIV *Congrès International d'Anthropologie et d'Archéologie Préhis-toriques* 1: 165–238.

Briggs, A. et al. 2009. "Targeted Retrieval and Analysis of Five Neandertal mtDNA Genomes." *Science* 325: 318–21.

Brooks, A. S. et al. 2005. "Projectile Technologies of the African MSA: Implications for Modern Human Origins." In *Transitions Before the Transition: Evolution and Stability in the Middle Paleolithic and Middle Stone Age*, edited by E. Hovers and S. Kuhn, 233–55. New York: Kluwer.

Broom, R. 1932. *The Mammal-Like Reptiles of South Africa and the Origins of Mammals*. London: Witherby.

Broom, R. 1934. "On The Fossil Remains Associated with *Australopithecus africanus*." *South African Journal of Science* 31: 471–80.

Broom, R. 1936. "A New Fossil Anthropoid Skull from Sterkfontein, near Krugersdorp, South Africa." *Nature* 138: 486–88.

Broom, R. 1938. "Pleistocene Anthropoid Apes of South Africa." *Nature* 142: 377–9.

Broom, R. 1949. "Another New Type of Fossil Ape-Man (*Paranthropus crassidens*)." *Nature* 163: 57.

Broom, R. 1950. *Finding the Missing Link*. London: Watts and Co.

Broom, R., and Robinson, J. T. 1949. "A New Type of Fossil Man." *Nature* 164: 322–23.

Broom, R., and Robinson, J. T. 1950. "Man Contemporaneous with the Swartkrans Ape-Man." *American Journal of Physical Anthropology* 8: 151–55.

Broom, R., and Robinson, J. T. 1952. *Swartkrans Ape-Man* Paranthropus crassidens. Pretoria: Transvaal Museum.

Broom, R., and Schepers, G. W. H. 1946. *The South African Fossil Ape-Men: The Australopithecinae*. Pretoria: Transvaal Museum.

Brown, F. H., McDougall, I., and Fleagle, J. G. 2012. "Correlation of the KHS Tuff of the Kibish Formation to Volcanic Ash Layers at Other Sites, and the Age of Early *Homo sapiens* (Omo I and II)." *Journal of Human Evolution*, in press.

Brown, F. H. et al. 1985. "Early *Homo erectus* Skeleton from West Lake Turkana, Kenya." *Nature* 316: 788–92.

Brown, K. S. et al. 2009. "Fire as an Engineering Tool of Early Modern Humans." *Science* 325: 859–62.

Brunet, M. et al. 2002. "A New Hominid from the Upper Miocene of Chad, Central Africa." *Nature* 418: 145–51.

Bunn, H. T. 1981. "Archaeological Evidence for Meat-Eating by Plio-Pleistocene Hominids from Koobi Fora and Olduvai Gorge." *Nature* 291: 574–77.

Bunn, H. T. 1982a. *Meat-Eating and Human Evolution: Studies on the Diet and Subsistence Patterns of Plio-Pleistocene Hominids in East Africa*. Ph.D. dissertation. Berkeley: University of California.

Bunn, H. T. 1982b. "Book Review of *Bones: Ancient Men and Modern Myths*, by L.R. Binford." *Science* 215: 494–95.

Bunn, H. T. 1986. "Patterns of Skeletal Part Representation and Hominid Subsistence Activities at Olduvai Gorge, Tanzania, and Koobi Fora, Kenya." *Journal of Human Evolution* 15: 673–90.

Bunn, H. T. 1991. "A Taphonomic Perspective on the Archaeology of Human Origins." *Annual Review of Anthropology* 20: 433–67.

Bunn, H. T. 2001. "Hunting, Power Scavenging, and Butchering by Hadza Foragers and by Plio-Pleistocene *Homo*." In *Meat-Eating and Human Evolution*, edited by C. B. Stanford and H. T. Bunn, 199–218. New York: Oxford University Press.

Bunn, H. T. 2007. "Meat Made Us Human." In *Evolution of the Human Diet: The Known, the Unknown, and the Unknowable*, edited by P. Ungar, 191–211. Oxford, UK: Oxford University Press.

Bunn, H. T., and Kroll, E. M. 1986. "Systematic Butchery by Plio/Pleistocene Hominids at Olduvai Gorge, Tanzania." *Current Anthropology* 27: 431–52.

Bunn, H. T., and Kroll, E. M. 1988. "Reply to Binford." *Current Anthropology* 29: 135–49.

Bunn, H. T., and Pickering, T. R. 2010a. "Bovid Mortality Profiles in Paleoecological Context Falsify Hypotheses of Endurance Running Hunting and Passive Scavenging by Early Pleistocene Hominins." *Quaternary Research* 74: 395–404.

Bunn, H. T., and Pickering, T. R. 2010b. "Methodological Recommendations for Ungulate Mortality Analyses in Paleoanthropology." *Quaternary Research* 74: 388–94.

Burbano, H. A. et al. 2010. "Targeted Investigation of the Neandertal Genome by Array-Based Sequence Capture." *Science* 328: 723–25.

Burkhardt, R. W. 2005. *Patterns of Behavior: Konrad Lorenz, Niko Tinbergen, and the Founding of Ethology*. Chicago: University of Chicago Press.

Burnett, J. 1773–92. *Of the Origin and Progress of Language*, 6 vols. Edinburgh: J. Balfour.

Campbell, H. 1904. "The Evolution of Man's Diet." *Lancet*: 781–84; 848–51; 909–12; 967–69; 1097–99; 1234–37; 1368–70; 1519–22; 1667–70.

Campbell, H. 1917. "The Biological Aspects of War." *Lancet*: 433–35; 469–71; 505–8.

Campbell, H. 1921. "Man's Evolution from Anthropoid." *Lancet* 2: 629.

Capaldo, S. D. 1997. "Experimental Determinations of Carcass Processing by Plio-Pleistocene Hominids and Carnivores at FLK 22 (*Zinjanthropus*), Olduvai Gorge, Tanzania." *Journal of Human Evolution* 33: 555–97.

Capaldo, S. D., and Blumenschine, R. J. 1994. "A Quantitative Diagnosis of Notches Made by Hammerstone Percussion and Carnivore Gnawing on Bovid Long Bones." *American Antiquity* 59: 724–48.

Capaldo, S. D., and Peters, C. R. 1995. "Skeletal Inventories from Wildebeest Drowning at Lakes Masek and Ndutu in the Serengeti Ecosystem of Tanzania." *Journal of Archaeological Science* 22: 385–408.

Carrier, D. R. 2007. "The Short Legs of Great Apes: Evidence for Aggressive Behavior in Australopiths." *Evolution* 61: 596–605.

Cartmill, M. 1993. *A View to a Death in the Morning: Hunting and Nature through History*. Cambridge, MA: Harvard University Press.

Cattelin, P. 1989. "Un crochet de propulser solutréen de la grotte de Combe-Saunière 1 (Dordogne)." *Bulletin de la Société Préhistorique Française* 86: 213–16.

Cattelin, P. 1997. "Hunting during the Upper Paleolithic: Bow, Spearthrower, or Both?" In *Projectile Technology*, edited by H. Knecht, 213–40. New York: Plenum Press.

Cavallo, J. A. 1997. *A Re-Examination of Isaac's Central-Place Foraging Hypothesis*. Ph.D. dissertation. New Brunswick, NJ: Rutgers University.

Cavallo, J. A., and Blumenschine, R. J. 1989. "Tree-Stored Leopard Kills: Expanding the Hominid Scavenging Niche." *Journal of Human Evolution* 18: 393–99.

Cerling, T. E. 1992. "Development of Grasslands and Savannas in East Africa during the Neogene." *Palaeogeography, Palaeoclimatology, Palaeoecology* 97: 241–47.

Cerling, T. E., Bowman, J. R., and O'Neil, J. R. 1988. "An Isotopic Study of a Fluvial-Lacustrine Sequence: The Plio-Pleistocene Koobi Fora Sequence." *Palaeogeography, Palaeoclimatology, Palaeoecology* 63: 335–56.

Cerling, T. E. et al. 2011a. "Woody Cover and Hominin Environments in the Last 6 Million Years." *Nature* 476: 51–56.

Cerling, T. E. et al. 2011b. "Diet of *Paranthropus boisei* in the Early Pleistocene of East Africa." *Proceedings of the National Academy of Science (USA)* 108: 9337–41.

Chazan, M. et al. 2008. "Radiometric Dating of the Earlier Stone Age Sequence in Excavation I at Wonderwerk Cave, South Africa." *Journal of Human Evolution* 55: 1–11.

Christensen, B., and Maslin, M. M. (editors). 2007. "Special Issue: African Paleoclimate and Human Evolution." *Journal of Human Evolution* 53: 443–634.

Churchill, S. E. 1993. "Weapon Technology, Prey Size Selection, and Hunting Methods in Modern Hunter-Gatherers: Implications for Hunting in the Palaeolithic and Mesolithic." In *Hunting and Animal Exploitation in the Later Palaeolithic and Mesolithic of Europe. American Anthropological Association Archaeological Papers*, edited by G. L. Peterkin, H. Bricker, and P. A. Mellars, 4: 11–24.

Churchill, S. E. 1998. "Cold Adaptation, Heterochrony, and the Neandertals." *Evolutionary Anthropology* 7: 46–61.

Churchill, S. E. 2006. "Bioenergetic Perspectives on Neandertal Thermoregulatory and Activity Budgets." In *Neandertals Revisited: New Approaches and Perspectives*, edited by K. Havarti and T. Harrison, 113–33. New York: Springer.

Churchill, S. E., Weaver, A. H., and Niewoehner, W. A. 1996. "Late Pleistocene Human Technological and Subsistence Behavior: Functional Interpretations of Upper Limb Morphology." In *Reduction Processes ("Chaines Opératoires") in the European Mousterian*, edited by A. Bietti and S. Grimaldi. *Quaternaria Nova* 6: 18–51.

Clark, J. D., and Harris, J. W. K. 1985. "Fire and Its Roles in Early Hominid Lifeways." *African Archaeological Review* 3: 3–27.

Clark, J. D. et al. 2003. "Stratigraphic, Chronological and Behavioural Contexts of Pleistocene *Homo sapiens* from Middle Awash, Ethiopia." *Nature* 423: 747–52.

Clarke, R. J. 1988. "A New *Australopithecus* Cranium from Sterkfontein and Its Bearing on the Ancestry of *Paranthropus*." In *Evolutionary History of the "Robust" Australopithecines*, edited by F. E. Grine, 285–92. New York: Aldine de Gruyter.

Clarke, R. J. 1994a. "The Significance of the Swartkrans *Homo* to the *Homo erectus* Problem." *Courier Forschungs-Institut Senckenberg* 171: 185–93.

Clarke, R. J. 1994b. "Advances in Understanding the Craniofacial Anatomy of South African Early Hominids." In *Integrative Paths to the Past: Essays in Honor of F. Clark Howell*, edited by R. S. Corruccini and R. L. Ciochon, 205–22. New Jersey: Prentice Hall.

Clarke, R. J. 1994c. "On Some New Interpretations of Sterkfontein Stratigraphy." *South African Journal of Science* 90: 211–14.

Clarke, R. J. 1998. "First Ever Discovery of a Well Preserved Skull and Associated Skeleton of *Australopithecus*." *South African Journal of Science* 94: 460–63.

Clarke, R. J. 1999. "Discovery of Complete Arm and Hand of the 3.3 Million Year Old *Australopithecus* Skeleton from Sterkfontein." *South African Journal of Science* 95: 477–80.

Clarke, R. J. 2002. "Newly Revealed Information on the Sterkfontein Member 2 *Australopithecus* Skeleton." *South African Journal of Science* 98: 523–26.

Clarke, R. J. 2007. "Taphonomy of Sterkfontein *Australopithecus* Skeletons." In *Breathing Life into Fossils: Taphonomic Studies in Honor of C. K. (Bob) Brain*, edited by T. R. Pickering, K. Schick, and N. Toth, 195–201. Bloomington, IN: Stone Age Institute Press.

Clarke, R. J. 2008. "Latest Information on Sterkfontein's *Australopithecus* Skeleton and a New Look at *Australopithecus*." *South African Journal of Science* 104: 443–49.

Clarke, R. J. 2012. "A *Homo habilis* Maxilla and Other Newly-Discovered Hominid Fossils from Olduvai Gorge, Tanzania." *Journal of Human Evolution* 63: 418–28.

Clarke, R. J. 2013. "*Australopithecus* from Sterkfontein Caves, South Africa." In *Paleobiology of* Australopithecus, edited by K. Reed, J. Fleagle, and R. E. F. Leakey. New York: Springer, in press.

Clarke, R. J., and Tobias, P. V. 1995. "Sterkfontein Member 2 Foot Bones of the Oldest South African Hominid." *Science* 269: 521–24.

Cleghorn, N., and Marean, C. W. 2007. "The Destruction of Skeletal Elements by Carnivores: The Growth of a General Model for Skeletal Element Destruction and Survival in Zooarchaeological Assemblages." In *Breathing Life into Fossils: Taphonomic Studies in Honor of C. K. (Bob) Brain*, edited by T. R. Pickering, K. Schick, and N. Toth, 37–66. Bloomington, IN: Stone Age Institute Press.

Clutton-Brock, T. H. 1985. "Size, Sexual Dimorphism and Polygamy in Primates." In *Size and Scaling in Primate Biology*, edited by W. L. Jungers, 51–60. New York: Plenum Press.

Collard, M., and Wood, B. A. 2007. "Defining the Genus *Homo.*" In *Handbook of Paleoanthropology, Vol. 3: Phylogeny of Hominids,* edited by W. Henke and I. Tattersall, 1575–1611. Berlin: Springer.

Conrad, J. 1911. *Under Western Eyes.* London: Methuen.

Constantino, P., and Wood, B. A. 2009. "The Evolution of *Zinjanthropus boisei.*" *Evolutionary Anthropology* 16: 49–62.

Constantino, P., and Wright, B. W. 2009. "The Importance of Fallback Foods in Primate Ecology and Evolution." *American Journal of Physical Anthropology* 140: 599–602.

Constantino, P. et al. 2010. "Tooth Chipping Can Reveal the Diet and Bite Forces of Fossils Hominins." *Biology Letters* 6: 826–29.

Cooke, H. B. S. 1991. "*Dinofelis barlowi* (Mammalia, Carnivora, Felidae) Cranial Material from Bolt's Farm, Collected by the University of California African Expedition." *Palaeontologia Africana* 28: 9–21.

Copeland, S. R. et al. 2011. "Strontium Isotope Evidence for Landscape Use by Early Hominins." *Nature* 474: 76–79.

Cowgill, L. W. 2007. "Humeral Torsion Revisited: A Functional and Ontogenetic Model for Populational Variation." *American Journal of Physical Anthropology* 134: 472–80.

Currie, P. 2004. "Muscling In on Hominid Evolution." *Nature* 428: 373–74.

Daly, M., and Wilson, M. 1988. *Homicide.* New York: Aldine de Gruyter.

D'Anastasio, R. et al. 2009. "Possible Brucellosis in an Early Hominin Skeleton from Sterkfontein, South Africa." *PLoS ONE* 4: e6439.

Dart, R. A. 1925. "*Australopithecus africanus:* The Man-Ape of South Africa." *Nature* 115: 195–99.

Dart, R. A. 1926. "Taungs and Its Significance." *Natural History* 26: 315–27.

Dart, R. A. 1929. "A Note on the Taungs Skull." *South African Journal of Science* 26: 648–58.

Dart, R. A. 1948. "The Makapansgat Protohuman *Australopithecus prometheus.*" *American Journal of Physical Anthropology* 6: 259–83.

Dart, R. A. 1949. "The Predatory Implemental Technique of *Australopithecus.*" American Journal of Physical Anthropology 7: 1–38.

Dart, R. A. 1953. "The Predatory Transition from Ape to Man." *International Anthropological and Linguistic Review* 1: 201–18.

Dart, R. A. 1956a. "Cultural Status of the South African Man-Apes." *Smithsonian Report* 4240: 317–38.

Dart, R. A. 1956b. "The Myth of the Bone-Accumulating Hyaena." *American Anthropologist* 58: 40–62.

Dart, R. A. 1957. *The Osteodontokeratic Culture of* Australopithecus prometheus. Pretoria: Transvaal Museum.

Dart, R. A. 1960. "The Bone Tool-Manufacturing Ability of *Australopithecus prometheus.*" *American Anthropologist* 62: 134–43.

Dart, R. A., and Craig, D. 1959. *Adventures with the Missing Link.* New York: Harper and Brothers.

Darwin, C. 1871. *The Descent of Man and Selection in Relation to Sex.* London: John Murray.

Dean, M. C. 2000. "Progress in Understanding Hominoid Dental Develop-ment." *Journal of Anatomy* 197: 77–101.

Dean, M. C. 2006. "Tooth Microstructure Tracks the Pace of Human Life History Evolution." *Proceedings of the Royal Society (London) Series B* 273: 2799–802.

Dean, M. C., and Smith, B. H. 2009. "Growth and Development in the Nariokotome Youth, KNM WT 15000." In *The First Humans: Origin and Early Evolution of the Genus* Homo, edited by F. E. Grine, J. G. Fleagle, and R. E. F. Leakey, 101–20. New York: Springer.

Dean, M. C. et al. 1993. "Histological Reconstruction of Dental Development and Age at Death of a Juvenile *Paranthropus robustus* Specimen, SK 63, from Swartkrans, South Africa." *American Journal of Physical Anthropology* 91: 401–19.

Dean, M. C. et al. 2001. "Growth Processes in Teeth Distinguish Modern Humans from *Homo erectus* and Earlier Hominins." *Nature* 414: 628–31.

Defleur, A. et al. 1993. "Cannibalism among Neandertals?" *Nature* 362: 214.

Defleur, A. et al. 1999. "Neandertal Cannibalism at Moula-Guercy, Ardèche, France." *Science* 286: 128–31.

de Heinzelin, J. et al. 1999. "Environment and Behavior of 2.5-Million-Year-Old Bouri Hominids." *Science* 284: 625–29.

Deichmann, U. 1999. *Biologists under Hitler*. Cambridge, MA: Harvard University Press.

Delisle, R. G. 2006. *Debating Humankind's Place in Nature, 1860–2000: The Nature of Paleoanthropology*. Upper Saddle River, NJ: Prentice Hall.

de Maistre, J.-M. 1836. *Les Soirees de Saint-Pétersbourg, Vol. II*. Lyon: Péagaud, Lesne et Crozet.

de Maistre, J.-M. (translated by J. Lively). 1965. *The Works of Joseph de Maistre*. New York: MacMillan.

DeSilva, J. M., Proctor, D., and Zipfel, B. 2012. "A Complete Second Metatarsal (StW 89) from Sterkfontein Member 4, South Africa." *Journal of Human Evolution* 63: 487–96.

de Waal, F., and Lanting, F. 1997. *Bonobo: The Forgotten Ape*. Berkeley: University of California Press.

Dirks, P. H. G. M. et al. 2010. "Geological Setting and Age of *Australopithecus sediba* from Southern Africa." *Science* 328: 205–8.

Domínguez-Rodrigo, M. 1999. "Flesh Availability and Bone Modifications in Carcasses Consumed by Lions: Palaeoecological Relevance in Hominid Foraging Patterns." *Palaeogeography, Palaeoclimatology, Palaeoecology* 149: 373–83.

Domínguez-Rodrigo, M. 2001. "A Study of Carnivore Competition in Riparian and Open Habitats of Modern Savannas and Its Implications for Hominid Behavioral Modeling." *Journal of Human Evolution* 40: 77–98.

Domínguez-Rodrigo, M. 2002. "Hunting and Scavenging by Early Humans: The State of the Debate." *Journal of World Prehistory* 16: 1–54.

Domínguez-Rodrigo, M., and Barba, R. 2006. "New Estimates of Tooth Mark and Percussion Mark Frequencies at the FLK Zinj Site: The

Carnivore-Hominid-Carnivore Hypothesis Falsified." *Journal of Human Evolution* 50: 170–94.

Domínguez-Rodrigo, M., Barba, R., and Egeland, C. P. 2007. *Deconstructing Olduvai: A Taphonomic Study of the Bed I Sites.* Dordrecht, The Netherlands: Springer.

Domínguez-Rodrigo, M., and Pickering, T. R. 2003. "Early Hominid Hunting and Scavenging: A Zooarchaeological Review." *Evolutionary Anthropology* 12: 275–82.

Domínguez-Rodrigo, M., Pickering, T. R., and Bunn, H. T. 2010. "Configurational Approach to Identifying the Earliest Hominin Butchers." *Proceedings of the National Academy of Sciences (USA)* 107: 20929–34.

Domínguez-Rodrigo, M., Pickering, T. R., and Bunn, H. T. 2011. "Doubting Dikika Is About Data, Not Paradigms." *Proceedings of the National Academy of Sciences (USA)* 108: E117.

Domínguez-Rodrigo, M., Pickering, T. R., and Bunn, H. T. 2012. "Experimental Study of Cut Marks Made with Rocks Unmodified by Human Flaking and Its Bearing on Claims of ~3.4-Million-Year-Old Butchery Evidence from Dikika, Ethiopia." *Journal of Archaeological Science* 39: 205–14.

Domínguez-Rodrigo, M. et al. 2001. "Woodworking Activities by Early Humans: A Plant Residue Analysis on Acheulian Stone Tools from Peninj (Tanzania)." *Journal of Human Evolution* 40: 289–99.

Domínguez-Rodrigo, M. et al. 2005. "Cut Marked Bones from Pliocene Archaeological Sites at Gona, Afar, Ethiopia: Implications for the Function of the World's Oldest Stone Tools." *Journal of Human Evolution* 48: 109–21.

Domínguez-Rodrigo, M. et al. 2009a. "Unraveling Hominin Behavior at Another Anthropogenic Site from Olduvai Gorge (Tanzania): New Archaeological and Taphonomic Research at BK, Upper Bed II." *Journal of Human Evolution* 57: 260–83.

Domínguez-Rodrigo, M. et al. 2009b. "A New Protocol to Differentiate Trampling Marks from Butchery Cut Marks." *Journal of Archaeological Science* 36: 2643–54.

Domínguez-Rodrigo, M. et al. (editors). 2010. "Special Issue: Paleoecology and Hominin Behavior during Bed I at Olduvai Gorge (Tanzania)." *Quaternary Research* 74: 301–424.

Domínguez-Rodrigo, M. et al. 2012. "Earliest Porotic Hyperostosis on a 1.5-Million-Year-Old Hominin from Olduvai Gorge, Tanzania." *PLoS ONE*, in press.

Dominy, N. et al. 2008. "Mechanical Properties of Plant Underground Storage Organs and Implications for Dietary Models of Early Hominins." *Evolutionary Biology* 35: 159–75.

Douglas, J. K. E. 1969. "Lime in South Africa." *Journal of the South African Institute of Mining and Metallurgy* (August): 13–24.

Dugard, J. 1995. "Palaeontologist Ron Clarke and the Discovery of 'Little Foot': A Contemporary History." *South African Journal of Science* 91: 563–66.

Dunsworth, H., and Walker, A. C. 2002. "Early Genus *Homo*." In *The Primate Fossil Record*, edited by W. C. Hartwig, 419–26. Cambridge, UK: Cambridge University Press.

Egeland, C. P. 2007. *Zooarchaeological and Taphonomic Perspectives on Hominid-Carnivore Interactions at Olduvai Gorge, Tanzania.* Ph.D. dissertation. Bloomington: Indiana University.

Egeland, C. P., and Domínguez-Rodrigo, M. 2008. "Taphonomic Perspectives on Hominin Site Use and Foraging Strategies during Bed II Times at Olduvai Gorge, Tanzania." *Journal of Human Evolution* 55: 1031–52.

Ellis, C. J. 1997. "Factors Influencing the Use of Stone Projectile Tips: An Ethnographic Perspective." In *Projectile Technology*, edited by H. Knecht, 37–74. New York: Plenum Press.

Ellroy, J. 1996. *My Dark Places.* New York: Knopf.

El-Najjar, M. Y., Lozoff, B., and Ryan, D. 1975. "The Paleoepidemiology of Porotic Hyperostosis in the American Southwest: Radiological and Ecological Considerations." *American Journal of Roentgenology* 125: 918–24.

Ember, C. R. 1978. "Myths about Hunter-Gatherers." *Ethnology* 17: 439–48.

Ewer, R. F. 1973. *The Carnivores.* London: Weidenfeld and Nicholson.

Fabre, P-H., Rodrigues, A., and Douzery, E. J. P. 2009. "Patterns of Macroevolution among Primates Inferred From a Supermatrix of Mitochondrial and Nuclear DNA." *Molecular Phylogenetics and Evolution* 53: 808–25.

Fernández-Jalvo, Y. et al. 1996. "Evidence of Early Cannibalism." *Science* 271: 277–78.

Fernández-Jalvo, Y. et al. 1999. "Human Cannibalism in the Early Pleistocene of Europe (Gran Dolina, Sierra de Atapuerca, Burgos, Spain)." *Journal of Human Evolution* 37: 591–622.

Ferring, R. et al. 2011. "Earliest Human Occupations at Dmanisi (Georgian Caucasus) Dated to 1.85–1.78 Ma." *Proceedings of the National Academy of Sciences (USA)* 108: 10432–36.

Findlay, G. H. 1972. *Dr. Robert Broom, Palaeontologist and Physician (1866–1951): Biography/Appreciation/Bibliography.* Cape Town: A. A. Balkema.

Fiorillo, A. R. 1989. "An Experimental Study of Trampling: Implications for the Fossil Record." In *Bone Modification*, edited by R. Bonnichsen and M. Sorg, 61–71. Orono, ME: Center for the Study of the First Americans.

Foger, B., and Taschwer, K. 2001. *Die andere Seite des Spiegels: Konrad Lorenz und der Nationalsozialismus.* Vienna: Czernin Verlag.

Fossey, D. 1983. *Gorillas in the Mist.* Boston: Houghton Mifflin.

Friedman, L. J. 2001. "Erick Erikson on Identity, Generativity, and Pseudospeciation: A Biographer's Perspective." *Psychoanalysis and History* 3: 179–92.

Frison, G. C. 1989. "Experimental Use of Clovis Weaponry and Tools on African Elephants." *American Antiquity* 54: 766–84.

Furuichi, T. 2006. "Evolution of the Social Structure of Hominoids: Reconsideration of Food Distribution and the Estrus Sex Ratio." In *Human Origins and Environmental Backgrounds*, edited by H. Ishida et al., 235–48. New York: Springer.

Furuichi, T. 2011. "Female Contributions to the Peaceful Nature of Bonobo Society." *Evolutionary Anthropology* 20: 131–42.

Gabunia, L., and Vekua, A. A. 1995. "A Plio-Pleistocene Hominid from Dmanisi, East Georgia, Caucasus." *Nature* 373: 509–12.

Gabunia, L. et al. 2000. "Earliest Pleistocene Hominid Cranial Remains from Dmanisi, Republic of Georgia." *Science* 288: 1019–25.

Galán, A. B. et al. 2009. "A New Experimental Study on Percussion Marks and Notches and Their Bearing on the Interpretation of Hammerstone-Broken Faunal Assemblages." *Journal of Archaeological Science* 36: 776–84.

Galdikas, B. M. F. 1996. *Reflections of Eden: My Years with the Orangutans of Borneo.* New York: Back Bay Books.

Gamble, C. 1986. *The Paleolithic Settlement of Europe.* Cambridge, UK: Cambridge University Press.

Gamble, C. 1987. "Man the Shoveler: Alternative Models for Paleolithic Colonization and Occupation in Northern Latitudes." In *The Pleistocene Old World: Regional Perspectives,* edited by O. Soffer, 81–98. New York: Plenum Press.

Gardiner, B. 2003. "The Piltdown Forgery: A Re-Statement of the Case against Hinton." *Zoological Journal of the Linnaean Society* 139: 315–35.

Gathogo, P. N., and Brown, F. H. 2006. "Revised Stratigraphy of Area 123, Koobi Fora, Kenya, and Age Estimates of Its Fossil Mammals, Including Hominins." *Journal of Human Evolution* 51: 471–79.

Germonpré, M., Lázničkova-Galetová, M., and Sablin, M. 2012. "Palaeolithic Dog Skulls at the Gravettian Předmosti Site, the Czech Republic." *Journal of Archaeological Science* 39: 184–202.

Germonpré, M. et al. 2009. "Fossil Dogs and Wolves from Palaeolithic Sites in Belgium, the Ukraine and Russia: Osteometry, Ancient DNA and Stable Isotopes." *Journal of Archaeological Science* 36: 473–90.

Germonpré, M. et al. 2012. "Paleolithic Dogs and the Early Domestication of the Wolf: A Reply to the Comments of Crockford and Kuzmin." *Journal of Archaeological Science,* in press.

Ghiglieri, M. P. 1987. "Socioecology of the Great Apes and the Hominid Ancestor." *Journal of Human Evolution* 16: 319–57.

Gibbons, A. 2007. *The First Human: The Race to Discover Our Earliest Ancestors.* New York: Anchor.

Gilbert, C. C., McGraw, W. S., and Delson, E. 2009. "Plio-Pleistocene Eagle Predation on Fossil Cercopithecids from the Humpata Plateau, Southern Angola." *American Journal of Physical Anthropology* 139: 421–29.

Gilby, I. C. et al. 2010. "No Evidence of Short-Term Exchange of Meat for Sex among Chimpanzees." *Journal of Human Evolution* 49: 44–53.

Gonzales, L. 2004. *Deep Survival: Who Lives, Who Dies and Why.* New York: Norton.

Goodall, J. 1964. "Tool-Using and Aimed Throwing in a Community of Free-Living Chimpanzees." *Nature* 201: 265–66.

Goodall, J. 1971. *In the Shadow of Man.* Boston: Houghton Mifflin.

Goodall, J. 1986. *The Chimpanzees of Gombe: Patterns of Behavior.* Cambridge, MA: Belknap Press.

Goodall, J. 1996. *My Life with Chimpanzees.* New York: Aladdin.

Goodall, J. 2000. *Through the Window.* New York: Mariner Books.

Goodman, M. 1963. "Man's Place in the Phylogeny of the Primates as Reflected in Serum Proteins." In *Classification and Human Evolution,* edited by S. L. Washburn, 204–34. Chicago: Aldine.

Goodman, M. 1998. "The Genomic Record of Humankind's Evolutionary Roots." *American Journal of Human Genetics* 64: 31–39.

Goren-Inbar, N. et al. 2004. "Evidence of Hominin Control of Fire at Gesher Benot Ya'aqov, Israel." *Science* 304: 725–27.

Gould, S. J. 1980. "The Piltdown Conspiracy." *Natural History* 89: 8–28.

Gould, S. J. 1981. "Piltdown in Letters." *Natural History* 90: 12–30.

Gowlett, J. A. J. et al. 1981. "Early Archaeological Sites, Further Hominid Remains and Traces of Fire from Chesowanja, Kenya." *Nature* 294: 125–29.

Graves, R. et al. 2010. "Just How Strapping Was KNM-WT 15000?" *Journal of Human Evolution* 59: 542–54.

Green, R. E. et al. 2006. "Analysis of One Million Base Pairs of Neandertal DNA." *Nature* 444: 330–36.

Green, R. E. et al. 2009. "The Neandertal Genome and Ancient DNA Authenticity." *The EMBO Journal* 28: 2494–502.

Green, R. E. et al. 2010. "A Draft Sequence of the Neandertal Genome." *Science* 328: 710–22.

Greenlee, D. M. 1996. "An Electron Microprobe Evaluation of Diagenetic Alteration in Archaeological Bone." In *Archaeological Chemistry*, edited by M. V. Oma, 334–54. Washington, DC: American Chemical Society Press.

Grine, F. E. (editor). 1988. *The Evolutionary History of the "Robust" Australopithecines*. New York: Aldine de Gruyter.

Grine, F. E., Fleagle, J. G., and Leakey, R. E. F. (editors). 2009. *The First Humans: Origin and Early Evolution of the Genus* Homo. New York. Springer.

Grine, F. E., Gwinnett, A. J., and Oaks, J. H. 1990. "Early Hominid Dental Pathology: Interproximal Caries in 1.5 Million Year Old *Paranthropus robustus* from Swartkrans." *Archives of Oral Biology* 35: 381–86.

Grine, F. E., and Kay, R. 1988. "Early Hominid Diets from Quantitative Image Analysis of Dental Microwear." *Nature* 333: 765–68.

Grine, F. E. et al. 2012. "The Enigmatic Molar from Gondolin, South Africa: Implications for *Paranthropus* Paleobiology." *Journal of Human Evolution*, in press.

Guatelli-Steinberg, D. 2003. "Macroscopic and Microscopic Analyses of Linear Enamel Hypoplasia in Plio-Pleistocene South African Hominins with Respect to Aspects of Enamel Development and Morphology." *American Journal of Physical Anthropology* 120: 309–22.

Guthrie, R. D. 2005. *The Nature of Paleolithic Art*. Chicago: University of Chicago Press.

Haile-Selassie, Y. 2001. "Late Miocene Hominids from the Middle Awash, Ethiopia." *Nature* 412: 178–81.

Haile-Selassie, Y., Suwa, G., and White, T. D. 2004. "Late Miocene Teeth from the Middle Awash, Ethiopia, and Early Hominid Dental Evolution." *Science* 303: 153–55.

Haile-Selassie, Y., and WoldeGabriel, G. (editors). 2009. Ardipithecus kadabba: *Late Miocene Evidence from the Middle Awash, Ethiopia*. Berkeley: University of California Press.

Haile-Selassie, Y. et al. 2010. "An Early *Australopithecus afarensis* Postcranium from Woranso-Mille, Ethiopia." *Proceedings of the National Academy of Science (USA)* 107: 12121–26.

Haile-Selassie, Y. et al. 2012. "A New Hominin Foot from Ethiopia Shows Multiple Pliocene Bipedal Adaptations." *Nature* 483: 565–70.

Hamilton, W. D. 1964. "The Genetical Evolution of Social Behavior, I and II." *Journal of Theoretical Biology* 7: 1–52.

Hammer, M. F. et al. 2001. "Hierarchical Patterns of Global Human Y-Chromosome Diversity." *Molecular Biology and Evolution* 18: 1189–203.

Haraway, D. 1988. "Remodelling the Human Way of Life: Sherwood Washburn and the New Physical Anthropology, 1950–1980." In *Bones, Bodies, Behavior: Essays on Biological Anthropology*, edited by G. Stocking, 206–59. Madison: University of Wisconsin Press.

Hardy, B. L. et al. 2001. "Stone Tool Function at the Paleolithic Sites of Starosele and Buran Kaya III, Crimea: Behavioral Implications." *Proceedings of the National Academy of Sciences (USA)* 98: 10972–77.

Hare, B. 2007. "From Nonhuman to Human Mind: What Changed and Why?" *Current Directions in Psychological Science* 16: 60–64.

Hare, B., and Tomasello, M. 2004. "Chimpanzees Are More Skillful in Competitive Than Cooperative Cognitive Tasks." *Animal Behaviour* 68: 571–81.

Hare, B., and Tomasello, M. 2005. "Human-Like Social Skills in Dogs?" *Trends in Cognitive Sciences* 9: 439–44.

Hare, B. et al. 2002. "The Domestication of Social Cognition in Dogs." *Science* 298: 1634–36.

Harris, S. 2010. *The Moral Landscape: How Science Can Determine Human Values*. New York: Free Press.

Harrison, T., and Rook, L. 1997. "Enigmatic Anthropoid or Misunderstood Ape? The Phylogenetic Status of *Oreopithecus bambolii* Reconsidered." In *Function, Phylogeny and Fossils: Miocene Hominoid Evolution and Adaptation*, edited by D. R. Begun, C. V. Ward, and M. D. Rose, 327–62. New York: Plenum Press.

Harrison Matthews, L. 1981. "Piltdown Man: The Missing Links." *New Scientist* 91: 26.

Hart, D., and Sussman, R. W. 2005. *Man the Hunted: Primates, Predators, and Human Evolution*. New York: Westview.

Hashimoto, C., and Furuichi, T. 2005. "Possible Intergroup Killing in Chimpanzees in the Kalinzu Forest, Uganda." *Pan African News* 12: 3–5.

Hawkes, K., O'Connell, J. F., and Blurton Jones, N. G. 2001. "Hadza Meat Sharing." *Evolution and Human Behavior* 22: 113–42.

Hay, R. 1976. *Geology of the Olduvai Gorge*. Berkeley: University of California Press.

Heaton, J. L., and Pickering, T. R. 2006. "Archaeological Analysis Does Not Support Intentionality in the Production of Brushed Ends on Chimpanzee Termiting Tools." *International Journal of Primatology* 27: 1619–33.

Hedenstrom, A. 1995. "Lifting the Taung Child." *Nature* 378: 670.

Hedges, C. 2003. *War Is a Force That Gives Us Meaning*. New York: Anchor.

Henry, A. G. et al. 2012. "The Diet of *Australopithecus sediba.*" Nature 487: 90–93.

Hernandez-Aguilar, R. A., Moore, J. J., and Pickering, T. R. 2007. "Savanna Chimpanzees Use Tools to Harvest the Underground Storage Organs of Plants." *Proceedings of the National Academy of Sciences (USA)* 104: 19210–13.

Hey, J. 2010. "The Divergence of Chimpanzee Species and Subspecies as Revealed in Multi-Population Isolation-with-Migration Analyses." *Molecular Biology and Evolution* 27: 921–33.

Hillson, S. 1996. *Dental Anthropology*. Cambridge, UK: Cambridge University Press.

Hiraiwa-Hasegawa, M. et al. 1986. "Aggression toward Large Carnivores by Wild Chimpanzees of Mahale Mountains National Park, Tanzania." *Folia Primatologica* 47: 8–13.

Hodgins, G., Brook, G. A., and Marais, E. 2007. "Bomb-Spike Dating of Mummified Baboon in Ludwig Cave, Namibia." *International Journal of Speleology* 36: 31–38.

Holloway, R. L., Broadfield, D. C., and Yuan, M. S. 2004. *The Human Fossil Record: Brain Endocasts*. Hoboken, NJ: Wiley-Liss.

Howell, F. C. 1978. "Hominidae." In *Evolution of African Mammals*, edited by V. Maglio and H. B. S. Cooke, 154–248. Cambridge, MA: Harvard University Press.

Huckell, B. B. 1979. "Of Chipped Stone Tools, Elephants, and the Clovis Hunters: An Experiment." *Plains Anthropologist* 24: 177–89.

Huckell, B. B. 1982. "The Denver Elephant Project: A Report on Experimentation and Thrusting Spears." *Plains Anthropologist* 27: 217–24.

Hughes, A. R. 1954. "Hyaenas versus Australopithecines as Agents of Bone Accumulations." *American Journal of Physical Anthropology* 12: 467–86.

Hughes, A. R., and Tobias, P. V. 1977. "A Fossil Skull Probably of the Genus *Homo* from Sterkfontein, Transvaal." *Nature* 265: 310–12.

Hume, D. 1772. *Essays and Treatises on Several Subjects, Vol. II: An Enquiry Concerning Human Understanding; A Dissertation on the Passions; An Enquiry Concerning the Principles of Morals; and The Natural History of Religion*. Edinburgh and London: T. Cadwell.

Hutton, J. 1788. "Theory of the Earth: Or an Investigation of the Laws Observable in the Composition, Dissolution, and Restoration of Land upon the Globe." *Transactions of the Royal Society of Edinburgh* 1: 209–304.

Hutton, J. 1794. *An Investigation of the Principles of Knowledge and of the Progress of Reason, from Sense to Science and Philosophy*. Edinburgh and London: A. Strahan and T. Cadwell.

Isaac, G. Ll. 1978. "The Food-Sharing Behavior of Protohuman Hominids." *Scientific American* 238: 90–108.

Isaac, G. Ll. 1981. "Stone Age Visiting Cards: Approaches to the Study of Early Land Use Patterns." In *Patterns of the Past*, edited by I. Hodder, G. Ll. Isaac, and N. Hammond, 205–57. Cambridge, UK: Cambridge University Press.

Isaac, G. Ll. 1984. "The Archaeology of Human Origins: Studies of the Lower Pleistocene in East Africa, 1971–1981." In *Advances in World Archaeology*,

*Vol. 3*, edited by F. Wendorf and A. E. Close, 1–87. Orlando, FL: Academic Press.

Isaac, G. Ll., and Curtis, G. H. 1974. "Age of Early Acheulean Industries from the Peninj Group, Tanzania." *Nature* 249: 624–27.

Itani, J. 1985. "Evolution of Primate Social Structure." *Man* 20: 593–611.

Johanson, D. C., Taieb, M., and Coppens, Y. (editors). 1982. "Special Issue: Pliocene Hominids from Hadar, Ethiopia." *American Journal of Physical Anthropology* 57: 373–719.

Jones, C., and Sabater Pí, J. 1969. "Sticks Used by Chimpanzees in Rio Muni, West Africa." *Nature* 223: 100–101.

Jones, P. R. 1981. "Experimental Implement Manufacture and Use: A Case Study from Olduvai Gorge." *Philosophical Transactions of the Royal Society (London) Series B* 292: 189–95.

Kalikow, T. 1983. "Konrad Lorenz's Ethological Theory: Explanation and Ideology, 1938–1943." *Journal of the History of Biology* 16: 39–73.

Kano, T. 1982. "The Social Group of Pygmy Chimpanzees (*Pan paniscus*) of Wamba." *Primates* 23: 171–88.

Kano, T. 1992. *The Last Ape: Pygmy Chimpanzee Behavior and Ecology*. Stanford, CA: Stanford University Press.

Katzenberg, M. A., Herring, D. A., and Saunders, S. R. 1996. "Weaning and Infant Mortality: Evaluating the Skeletal Evidence." *Yearbook of Physical Anthropology* 39: 177–99.

Keeley, L. H. 1996. *War before Civilization: The Myth of the Peaceful Savage*. New York: Oxford University Press.

Keeley, L. H., and Toth, N. 1981. "Microwear Polishes on Early Stone Tools from Koobi Fora, Kenya." *Nature* 293: 464–65.

Keyser, A. W. 2000. "The Drimolen Skull: The Most Complete Australopithecine Cranium and Mandible to Date." *South African Journal of Science* 96: 189–93.

Kibii, J. M. 2007. "Taxonomy, Taphonomy and Palaeoenvironment of Hominid and Non-Hominid Primates from the Jacovec Cavern, Sterkfontein." *South African Archaeological Bulletin* 62: 90–97.

Kimbel, W. H. 2009. "The Origin of *Homo*." In *The First Humans: Origin and Early Evolution of the Genus* Homo, edited by F. E. Grine, J. G. Fleagle, and R. E. F. Leakey, 31–37. New York: Springer.

Kimbel, W. H., Johanson, D. C., and Rak, Y. 1997. "Systematic Assessment of a Maxilla of *Homo* from Hadar, Ethiopia." *American Journal of Physical Anthropology* 103: 235–62.

Kimbel, W. H., Rak, Y., and Johanson, D. C. 2004. *The Skull of* Australopithecus afarensis. Oxford, UK: Oxford University Press.

Kimbel, W. H. et al. 1996. "Late Pliocene *Homo* and Oldowan Tools from the Hadar Formation (Kada Hadar Member), Ethiopia." *Journal of Human Evolution* 31: 549–61.

Klein, R. G. 1978. "Stone Age Predation on Large African Bovids." *Journal of Archaeological Science* 5: 195–217.

Klein, R. G. 1999. *The Human Career: Human Biological and Cultural Origins*, 2nd ed. Chicago: University of Chicago Press.

Knauft, B. B. 1987. "Reconsidering Violence in Simple Human Societies: Homicide among the Gebusi of New Guinea." *Current Anthropology* 28: 457–500.

Knauft, B. B. 1991. "Violence and Sociality in Human Evolution." *Current Anthropology* 32: 391–428.

Knecht, H. (editor). 1997. *Projectile Technology*. New York: Plenum Press.

Knight, A. et al. 2003. "African Y-Chromosome and mtDNA Divergence Provides Insight into the History of Click Languages." *Current Biology* 13: 464–73.

Kohler, M., and Moya Sola, S. J. 1997. "Ape-Like or Hominid-Like? The Positional Behavior of *Oreopithecus bambolii* Reconsidered." *Proceedings of the National Academy of Science (USA)* 94: 11747–50.

Kruuk, H. 1972. *The Spotted Hyaena: A Study of Predation and Social Behavior*. Chicago: University of Chicago Press.

Kuman, K., and Clarke, R. J. 2000. "Stratigraphy, Artifact Industries and Hominid Associations for Sterkfontein Member 5." *Journal of Human Evolution* 38: 827–47.

Kuykendall, K. L. 2003. "Reconstructing Australopithecine Growth and Development: What Do We Think We Know?" In *Patterns of Growth and Development in the Genus* Homo, edited by J. L. Thompson, G. E. Krovitz, and A. J. Nelson, 191–218. Cambridge, UK: Cambridge University Press.

Lacruz, R. S., Ramirez Rozzi, F., and Bromage, T. G. 2006. "Variation in Enamel Development of South African Fossil Hominids." *Journal of Human Evolution* 51: 580–90.

Lacruz, R. et al. 2008. "Patterns of Enamel Secretion in Fossil Hominins." *Journal of Anatomy* 213: 148–58.

Laden, G., and Wrangham, R. 2005. "The Rise of the Hominids as an Adaptive Shift in Fallback Foods: Plant Underground Storage Organs (USOs) and Australopith Origins." *Journal of Human Evolution* 49: 482–98.

Lallo, J. W., Armelagos, G. J., and Mensforth, R. P. 1977. "The Role of Diet, Disease, and Physiology in the Origin of Porotic Hyperostosis." *Human Biology* 49: 471–83.

Lamb, M. E., and Hewlett, B. S. (editors). 2005. *Hunter-Gatherer Childhoods: Developmental and Cultural Perspectives*. Piscataway, NJ: Aldine Transaction.

Larson, S. G. 2007. "Evolutionary Transformation of the Hominin Shoulder." *Evolutionary Anthropology* 16: 172–87.

Larson, S. G. 2009. "Evolution of the Hominin Shoulder: Early *Homo*." In *The First Humans: Origin and Early Evolution of the Genus* Homo, edited by F. E. Grine, J. G. Fleagle, and R. E. F. Leakey, 65–76. Dordrecht, The Netherlands: Springer.

Latimer, B. M., and Lovejoy, C. O. 1982. "Hominid Tarsal, Metatarsal, and Phalangeal Bones Recovered from the Hadar Formation: 1974–1977 Collections." *American Journal of Physical Anthropology* 57: 701–19.

Latimer, B. M., and Lovejoy, C. O. 1990a. "Hallucal Tarsometatarsal Joint in *Australopithecus afarensis*." *American Journal of Physical Anthropology* 82: 125–33.

Latimer, B. M., and Lovejoy, C. O. 1990b. "Metatarsophalangeal Joints of *Australopithecus afarensis.*" *American Journal of Physical Anthropology* 83: 13–23.

Latimer, B. M., and Ohman, J. C. 2001. "Axial Dysplasia in *Homo erectus.*" *Journal of Human Evolution* 40: A12.

Latimer, B. M., White, T. D., and Kimbel, W. H. 1981. "The Pygmy Chimpanzee Is Not a Living Missing Link in Human Evolution." *Journal of Human Evolution* 10: 475–88.

Lazuén, T. 2012. "European Neanderthal Stone Hunting Weapons Reveal Complex Behaviour Long Before the Appearance of Modern Humans." *Journal of Archaeological Science*, in press.

Leakey, L. S. B. 1934. *Adam's Ancestors: The Evolution of Man and His Culture.* London: Metheun.

Leakey, L. S. B. 1937. *White African.* London: Hodder and Stoughton.

Leakey, L. S. B. 1965. *Olduvai Gorge 1951–1961: A Preliminary Report on the Geology and Fauna.* Cambridge, UK: Cambridge University Press.

Leakey, L. S. B., Tobias, P. V., and Napier, J. R. 1964. "A New Species of the Genus *Homo* from Olduvai Gorge, Tanzania." *Nature* 210: 462–66.

Leakey, M. D. 1971. *Olduvai Gorge, Vol. 3: Excavations in Bed I and II, 1960–1963.* Cambridge, UK: Cambridge University Press.

Leakey, M. D. 1981. "Tracks and Tools." *Philosophical Transactions of the Royal Society (London) Series B* 292: 95–102.

Leakey, M. D. 1984. *Disclosing the Past.* London: Wiedenfeld and Nicolson.

Leakey, M. D., and Hay, R. L. 1979. "Pliocene Footprints in the Laetoli Beds at Laetoli, Northern Tanzania." *Nature* 278: 317–23.

Leakey, M. G. et al. 1995. "New Four Million-Year-Old Hominid Species from Kanapoi and Allia Bay, Kenya." *Nature* 376: 565–71.

Leakey, M. G. et al. 1998. "New Specimens and Confirmation of an Early Age for *Australopithecus anamensis.*" *Nature* 393: 62–66.

Leakey, R. E. F. 1983. *One Life: An Autobiography.* London: Michael Joseph.

LeBlanc, S. A. 2003. *Constant Battles: The Myth of the Peaceful, Noble Savage.* New York: St. Martin's.

Lee, R. B. 1979. *The !Kung San: Men, Women, and Work in a Foraging Society.* Cambridge, UK: Cambridge University Press.

Lee, R. B. 1982. "Politics, Sexual and Non-sexual, in an Egalitarian Society." In *Politics and History in Band Societies*, edited by E. Leacock and R. B. Lee, 37–59. Cambridge, UK: Cambridge University Press.

Lee, R. B., and Daly, R. 2002. *The Cambridge Encyclopedia of Hunters and Gatherers.* Cambridge, UK: Cambridge University Press.

Lee, R. B., and DeVore, I. (editors). 1968. *Man the Hunter.* Chicago: Aldine.

Lee-Thorp, J. A., and Sponheimer, M. 2006. "Contributions of Biogeochemistry to Understanding Hominin Dietary Ecology." *American Journal of Physical Anthropology* 131: 131–48.

Leigh, S. R. 1997. "Socioecology and the Ontogeny of Sexual Size Dimorphism in Anthropoid Primates." *American Journal of Physical Anthropology* 97: 339–56.

Lepre, C. J. et al. 2011. "An Earlier Origin for the Acheulian." *Nature* 477: 82–85.

Levin, N. E. et al. 2008. "Herbivore Enamel Carbon Isotopic Composition and the Environmental Context of *Ardipithecus* at Gona, Ethiopia." *Geological Society of American Special Paper* 446: 215–34.

Lewin, R. 1997. *Bones of Contention: Controversies in the Search for Human Origins.* Chicago: University of Chicago Press.

Lewis, A. B. 1991. "Comparisons between Dental and Skeletal Ages." *Angle Orthodontist* 61: 87–92.

Liebenberg, L. 2006. "Persistence Hunting by Modern Hunter-Gatherers." *Current Anthropology* 47: 1017–25.

Liem, K. F. 1980. "Adaptive Significance of Intra- and Interspecific Differences in the Feeding Repertoires of Cichlid Fishes." *American Zoologist* 20: 295–314.

Lindblad-Toh, K. et al. 2005. "Genome Sequence, Comparative Analysis and Haplotype Structure of the Domestic Dog." *Nature* 438: 803–19.

Liversidge, H. M. 2008. "Dental Age Revisited." In *Technique and Application in Dental Anthropology*, edited by J. D. Irish and G. C. Nelson, 234–52. Cambridge, UK: Cambridge University Press.

Lockwood, C. A. 1999. "Sexual Dimorphism in the Face of *Australopithecus africanus.*" *American Journal of Physical Anthropology* 108: 97–127.

Lockwood, C. A., and Tobias, P. V. 1999. "A Large Male Hominin Cranium from Sterkfontein, South Africa, and the Status of *Australopithecus africanus.*" *Journal of Human Evolution* 36: 637–85.

Lockwood, C. A., and Tobias, P. V. 2002. "Morphology and Affinities of New Hominin Cranial Remains from Member 4 of the Sterkfontein Formation, Gauteng Province, South Africa." *Journal of Human Evolution* 42: 389–450.

Lockwood, C. A. et al. 2007. "Extended Male Growth in a Fossil Hominin Species." *Science* 318: 1443–46.

Lordkipanidze, D. et al. 2005. "The Earliest Toothless Hominin Skull." *Nature* 434: 717–78.

Lordkipanidze, D. et al. 2006. "A Fourth Hominin Skull from Dmanisi, Georgia." *Anatomical Record* 288A: 1146–57.

Lordkipanidze, D. et al. 2007. "Postcranial Evidence for Early *Homo* from Dmanisi, Georgia." *Nature* 449: 305–10.

Lorenz, K. 1952. *King Solomon's Ring.* New York: Thomas Y. Crowell.

Lorenz, K. 1966. *On Aggression.* New York: Harcourt, Brace and World.

Lorenz, K. 1974. *Civilized Man's Eight Deadly Sins.* New York: Harcourt, Brace, Jovanovich.

Lorenz, K. 1979. *The Year of the Greylag Goose.* New York: Harcourt, Brace, Jovanovich.

Lorenz, K. 1981. *The Foundations of Ethology.* New York: Springer-Verlag.

Louchart, A. et al. 2009. "Taphonomic, Avian and Small-Vertebrate Indicators of *Ardipithecus ramidus.*" *Science* 326: 66e1–e4.

Lovejoy, C. O. 1981. "The Origin of Man." *Science* 211: 341–50.

Lovejoy, C. O. 2009. "Reexamining Human Origins in Light of *Ardipithecus ramidus.*" *Science* 326: 74e1–e8.

Lovejoy, C. O. et al. 2009a. "Careful Climbing in the Miocene: The Forelimbs of *Ardipithecus ramidus* and Humans Are Primitive." *Science* 326: 70e1–e8.

Lovejoy, C. O. et al. 2009b. "The Pelvis and Femur of *Ardipithecus ramidus:* The Emergence of Upright Walking." *Science* 326: 71e1–e6.

Lovejoy, C. O. et al. 2009c. "Combining Prehension and Propulsion: The Foot of *Ardipithecus ramidus.*" *Science* 326: 72e1–e8.

Lovejoy, C. O. et al. 2009d. "The Great Divides: *Ardipithecus ramidus* Reveals the Postcrania of Our Last Common Ancestors with African Apes." *Science* 326: 100–106.

Lucas, P. W. 2004. *Dental Functional Morphology: How Teeth Work.* Cambridge, UK: Cambridge University Press.

Lucas P. W. et al. 2008. "Dental Enamel as a Dietary Indicator in Mammals." *BioEssays* 30: 374–85.

Lyell, C. 1830–33. *Principles of Geology, or, The Modern Changes of the Earth and Its Inhabitants Considered Illustrative of Geology.* London: John Murray.

Lyman, R. L. 1994. *Vertebrate Taphonomy.* Cambridge, UK: Cambridge University Press.

MacLarnon, A. 1993. "The Vertebral Canal." In *The Nariokotome* Homo erectus *Skeleton,* edited by A. Walker and R. E. F. Leakey, 359–90. Cambridge, MA: Harvard University Press.

MacLarnon, A., and Hewitt, G. 1999. "The Evolution of Human Speech: The Role of Enhanced Breathing Control." *American Journal of Physical Anthropology* 109: 341–63.

MacLarnon, A., and Hewitt, G. 2004. "Increased Breathing Control: Another Factor in the Evolution of Human Language." *Evolutionary Anthropology* 13: 181–97.

Marchiafauai, A., Bonucci, E., and Ascenzi, A. 1974. "Fungal Osteoclasia: A Model of Dead Bone Resorption." *Calcified Tissues Research* 14: 195–210.

Marean, C. W. 1989. "Sabertooth Cats and Their Relevance for Early Hominid Diet and Evolution." *Journal of Human Evolution* 18: 559–82.

Marean, C. W., and Erhardt, C. L. 1995. "Paleoanthropological and Paleoecological Implications of the Taphonomy of a Sabertooth's Den." *Journal of Human Evolution* 29: 515–47.

Marlowe, F. W. 2010. *The Hadza: Hunter-Gatherers of Tanzania.* Berkeley: University of California Press.

Marlowe, F. W., and Berbesque, J. C. 2009. "Tubers as Fallback Foods and Their Impact on Hadza Hunter-Gatherers." *American Journal of Physical Anthropology* 140: 751–58.

Marshall, A., and Wrangham, R. W. 2007. "Evolutionary Consequences of Fallback Foods." *International Journal of Primatology* 28: 1219–35.

Marshall Thomas, E. 1959. *The Harmless People.* New York: Vintage Press.

Martin, D. L., and Frayer, D. W. (editors). 1997. *Troubled Times: Violence and Warfare in the Past.* Amsterdam: Gordon & Breach.

Mayr, E. 1950. "Taxonomic Categories in Fossil Hominids." *Cold Springs Harbor Symposium in Quantitative Biology* 15: 108–18.

McCarthy, C. 1985. *Blood Meridian, or the Evening Redness in the West.* New York: Vintage Books.

McCarthy, T., and Rubidge, B. 2006. *The Story of Earth and Life: A Southern African Perspective on a 4.6-Billion-Year Journey.* Cape Town: Struik.

McCollum, M. A. 1999. "The Robust Australopithecine Face: A Morphogenetic Perspective." *Science* 284: 301–5.

McCollum, M. A. et al. 2006. "Of Muscle-Bound Crania and Human Brain Evolution: The Story behind the MYH16 Headlines." *Journal of Human Evolution* 50: 232–36.

McDougall, C. 2009. *Born to Run: A Hidden Tribe, Superathletes, and the Greatest Race the World Has Ever Seen.* New York: Knopf.

McDougall, I., Brown, F. H., and Fleagle, J. G. 2005. "Stratigraphic Placement and Age of Modern Humans from Kibish, Ethiopia." *Nature* 433: 733–35.

McDougall, I., Brown, F. H., and Fleagle, J. G. 2008. "Sapropels and the Age of Hominins Omo I and II, Kibish, Ethiopia." *Journal of Human Evolution* 55: 409–20.

McGraw, W. S., Cooke, C., and Shultz, S. 2006. "Primate Remains from African Crowned Eagle (*Stephanoaetus coronatus*) Nests in Ivory Coast's Taï Forest: Implications for Primate Predation and Early Hominid Taphonomy in South Africa." *American Journal of Physical Anthropology* 131: 151–65.

McGrew, W. C. 1992. *Chimpanzee Material Culture: Implications for Human Evolution.* Cambridge, UK: Cambridge University Press.

McGrew, W. C. 2004. *The Cultured Chimpanzee: Reflections on Cultural Primatology.* Cambridge, UK: Cambridge University Press.

McGrew, W. C., Tutin, C. E. G., and Baldwin, P. J. 1979. "Chimpanzees, Tools and Termites: Cross-Cultural Comparisons of Senegal, Tanzania and Rio Muni." *Man* 14: 185–214.

McHenry, H. M. 1994. "Sexual Dimorphism in Fossil Hominids and Its Sociological Implications." In *Power, Sex and Tradition: The Archaeology of Human Ancestry*, edited by S. Shennan and J. Steele, 91–109. London: Routledge and Kegan Paul.

McKee, J. K. 1993. "The Formation and Geomorphology of Caves in Calcareous Tufas and Implications for the Study of the Taung Fossil Deposits." *Transactions of the Royal Society of South Africa* 48: 307–22.

McPherron, S. P. et al. 2010. "Evidence for Stone-Tool-Assisted Consumption of Animal Tissues before 3.39 Million Years Ago at Dikika, Ethiopia." *Nature* 466: 857–60.

McPherron, S. P. et al. 2011. "Tool-Marked Bones Prior to the Oldowan Change the Paradigm." *Proceedings of the National Academy of Sciences (USA)* 108: E116.

Menter, C. G. et al. 1999. "First Record of Hominid Teeth from the Plio-Pleistocene Site of Gondolin, South Africa." *Journal of Human Evolution* 37: 299–307.

Mercader, J., Panger, M., and Boesch, C. 2002. "Excavation of a Chimpanzee Stone Tool Site in the African Rainforest." *Science* 296: 1452–55.

Mercader, J. et al. 2007. "4,300-Year-Old Chimpanzee Sites and the Origins of Percussive Stone Technology." *Proceedings of the National Academy of Science (USA)* 104: 3043–48.

Mills, M. G. L. 1990. *Kalahari Hyaenas: Comparative Behavioural Ecology of Two Species*. London: Unwin Hyman.

Milton, K. 1999. "A Hypothesis to Explain the Role of Meat-Eating in Human Evolution." *Evolutionary Anthropology* 8: 11–21.

Milton, K. 2003. "The Critical Role Played by Animal Source Foods in Human Evolution." *Journal of Nutrition* 133: 3893–97.

Mitani, J. C., and Watts, D. 2001. "Why Do Chimpanzees Hunt and Share Meat?" *Animal Behaviour* 61: 915–24.

Moggi-Cecchi, J., Tobias, P. V., and Beynon, A. D. 1998. "The Mixed Dentition and Associated Fragments of a Juvenile Fossil Hominid from Sterkfontein, South Africa." *American Journal of Physical Anthropology* 106: 425–65.

Mokokowe, W. 2005. *Goldsmith's: A Preliminary Study of a Newly Discovered Pleistocene Site Near Sterkfontein*. M.Sc. thesis. Johannesburg: University of the Witwatersrand.

Molnar, S., and Molnar, I. M. 1985. "The Incidence of Enamel Hypoplasia among the Krapina Neandertals." *American Anthropologist* 87: 536–49.

Monahan, C. M. 1996. "New Zooarchaeological Data from Bed II, Olduvai Gorge, Tanzania: Implications for Hominin Behavior in the Early Pleistocene." *Journal of Human Evolution* 31: 93–128.

Morell, V. 1995. *Ancestral Passions: The Leakey Family and the Quest for Humankind's Beginnings*. New York: Simon and Schuster.

Morris, C. 1890. "From Brute to Man." *American Naturalist* 24: 341–50.

Movius, H. L. 1950. "A Wooden Spear of Third Interglacial Age from Lower Saxony." *Southwestern Journal of Anthropology* 6: 139–42.

Murray, S. et al. 2001. "Nutritional Composition of Some Wild Plant Foods and Honey Used by Hadza Foragers of Tanzania." *Journal of Food Composition and Analysis* 14: 3–13.

Naidoo, T. et al. 2010. "Development of a Single Base Extension Method to Resolve Y-Chromosome Haplogroups in Sub-Saharan African Populations." *Investigative Genetics* 1: 6.

Navarrete, A., van Schaik, C. P., and Isler, K. 2011. "Energetics and the Evolution of Human Brain Size." *Nature* 480: 91–93.

Newton-Fisher, N. E. 2007. "Chimpanzee Hunting Behaviour." In *Handbook of Paleoanthropology*, edited by W. Henke and I. Tattersall, 1295–320. Berlin: Springer.

Nishida, T. 1968. "The Social Group of Wild Chimpanzees in the Mahali Mountains." *Primates* 9: 167–201.

Nishida, T. et al. 1985. "Group Extinction and Female Transfer in Wild Chimpanzees in the Mahale National Park, Tanzania." *Zietschrift für Tierpsychologie* 67: 284–301.

Noonan, J. P. et al. 2006. "Sequencing and Analysis of Neandertal Genomic DNA." *Science* 314: 1113–18.

Oakley, K. P. 1954. "The Dating of the Australopithecinae of Africa." *American Journal of Physical Anthropology* 12: 9–28.

Oakley, K. P. 1956. "The Earliest Fire-Makers." *Antiquity* 30: 102–7.

Oakley, K. P. et al. 1977. "A Reappraisal of the Clacton Spearpoint." *Proceedings of the Prehistoric Society* 43: 13–30.

Odell, G. H., and Cowan, F. 1986. "Experiments with Spears and Arrows on Animal Targets." *Journal of Field Archaeology* 13: 195–212.

Ogden, J. A. 1990. "Histogenesis of the Musculoskeletal System." In *Nutrition and Bone Development*, edited by D. Simmons, 3–36. New York: Oxford University Press.

Ohman, J. C. et al. 2002. "Stature-at-Death of KNM-WT 15000." *Human Evolution* 17: 129–41.

Oliver, J. S. 1989. "Analogues and Site Context: Bone Damages from Shield Trap Cave (24CB91), Carbon County, Montana, USA." In *Bone Modification*, edited by R. Bonnichsen and M. Sorg, 73–98. Orono, ME: Center for the Study of the First Americans.

Olsen, S. L., and Shipman, P. 1988. "Surface Modifications on Bone: Trampling versus Butchery." *Journal of Archaeological Science* 15: 535–53.

Ortner, D. (editor). 2003. *Identification of Pathological Conditions in Human Skeletal Remains*. London: Academic Press.

Oxnard, C. 1975. *Uniqueness and Diversity in Human Evolution: Morphometric Studies of the Australopithecines*. Chicago: University of Chicago.

Oxnard, C. 1984. *The Order of Man: A Biomathematical Anatomy of the Primates*. New Haven: Yale University.

Palkovich, A. M. 1987. "Endemic Disease Patterns in Paleopathology: Porotic Hyperostosis." *American Journal of Physical Anthropology* 74: 527–37.

Panter-Brick, C., Layton, R. H., and Rowley-Conwy, P. 2001. *Hunter-Gatherers: An Interdisciplinary Perspective*. Cambridge, UK: Cambridge University Press.

Partridge, T. C. et al. 1999. "The New Hominid Skeleton from Sterkfontein, South Africa: Age and Preliminary Assessment." *Journal of Quaternary Science* 14: 293–98.

Pérez, P.-J. et al. 1997. "Pathological Evidence of the Cranial Remains from the Sima de los Huesos Middle Pleistocene Site (Sierra de Atapuerca, Spain): Description and Preliminary Inferences." *Journal of Human Evolution* 33: 409–21.

Peterson, D. 2008. *Jane Goodall: The Woman who Redefined Man*. New York: Mariner Books.

Peyton, J. 2001. *Solly Zuckerman: A Scientist Out of the Ordinary*. London: John Murray.

Pickering, R. et al. 2011. "*Australopithecus sediba* at 1.977 Ma and Implications for the Origins of the Genus *Homo*." *Science* 333: 1421–23.

Pickering, T. R. 1999. *Taphonomic Interpretations of the Sterkfontein Early Hominid Site (Gauteng, South Africa) Reconsidered in Light of Recent Evidence*. Ph.D. dissertation. Madison: University of Wisconsin.

Pickering, T. R. 2002. "Reconsideration of Criteria for Differentiating Faunal Assemblages Created by Hyenas and Hominids." *International Journal of Osteoarchaeology* 12: 127–41.

Pickering, T. R. 2006. "Subsistence Behaviour of South African Pleistocene Hominids." *South African Journal of Science* 102: 205–10.

Pickering, T. R., and Bunn, H. T. 2007. "The Endurance Running Hypothesis and Hunting and Scavenging in Savanna-Woodlands." *Journal of Human Evolution* 53: 438–42.

Pickering, T. R., and Bunn, H. T. 2012. "Another Take on Meat-Foraging by Pleistocene African Hominins: Tracking Behavioral Evolution beyond Baseline Inferences of Early Access to Carcasses." In *Stone Tools and Fossil Bones: Theoretical Debates in the Archaeology of Human Origins*, edited by M. Domínguez-Rodrigo, 152–73. Cambridge, UK: Cambridge University Press.

Pickering, T. R., Clarke, R. J., and Heaton, J. L. 2004. "The Context of StW 573, an Early Hominid Skull and Skeleton from Sterkfontein Member 2: Taphonomy and Paleoenvironment." *Journal of Human Evolution* 46: 277–95.

Pickering, T. R., Clarke, R. J., and Moggi-Cecchi, J. 2003. "Role of Carnivores in the Accumulation of the Sterkfontein Member 4 Hominid Assemblage: A Taphonomic Reassessment of the Complete Hominid Sample (1936–1999)." *American Journal of Physical Anthropology* 125: 1–15.

Pickering, T. R., and Domínguez-Rodrigo, M. 2006. "The Acquisition and Use of Large Mammal Carcasses by Oldowan Hominins in Eastern and Southern Africa: A Selected Review and Assessement." In *The Oldowan: Studies into the Origins of Human Technology*, edited by N. Toth and K. D. Schick, 113–28. Bloomington, IN: Stone Age Institute Press.

Pickering, T. R., and Domínguez-Rodrigo, M. 2010. "Chimpanzee Referents and the Emergence of Human Hunting." *Open Anthropology Journal* 3: 107–13.

Pickering, T. R., and Domínguez-Rodrigo, M. 2012. "Can We Use Chimpanzee Behavior to Model Early Hominin Hunting?" In *Stone Tools and Fossil Bones: Debates in the Archaeology of Human Origins*, edited by M. Domínguez-Rodrigo, 174–97. Cambridge, UK: Cambridge University Press.

Pickering, T. R., and Egeland, C. P. 2006. "Experimental Patterns of Hammerstone Percussion Damage on Bones: Implications for Inferences of Carcass Processing by Humans." *Journal of Archaeological Science* 33: 459–69.

Pickering, T. R., and Egeland, C. P. 2009. "Experimental Zooarchaeology and Its Role in Defining the Investigative Parameters of the Behavior of Early Stone Age Hominids." In *The Cutting Edge: New Approaches to the Archaeology of Human Origins*, edited by N. Toth and K. Schick, 171–77. Bloomington, IN: Stone Age Institute Press.

Pickering, T. R., and Heaton, J. L. 2009. "Roots, Bugs and Venison: Prehistoric Cuisine at Swartkrans Cave." *Quest* 5: 3–9.

Pickering, T. R., Schick, K., and Toth, N. (editors). 2007. *Breathing Life into Fossils: Taphonomic Studies in Honor of C. K. (Bob) Brain*. Bloomington, IN: Stone Age Institute Press.

Pickering, T. R., White, T. D., and Toth, N. 2000. "Cut Marks on a Plio-Pleistocene Hominid from Sterkfontein, South Africa." *American Journal of Physical Anthropology* 111: 579–84.

Pickering, T. R. et al. 2012. "New Hominid Fossils from Member 1 of the Swartkrans Formation, South Africa." *Journal of Human Evolution* 62: 618–28.

Pickering, T. R. et al. 2013. "New Taphonomic Diagnostic of Hominoid Behavior and the Consumption of Meat and Bone by 1.2-Million-Year-Old

Hominins at Olduvai Gorge, Tanzania." *Journal of Archaeological Science*, in press.

Pickford, M., and Senut, B. 2001. "The Geological and Faunal Context of Late Miocene Hominid Remains from Lukeino, Kenya." *Comptes Rendus de l'Académie de Sciences* 332: 145–52.

Pinker, S. 2002. *Blank Slate: The Modern Denial of Human Nature*. New York: Penguin.

Plavcan, J. M. 2000. "Inferring Social Behavior from Sexual Dimorphism in the Fossil Record." *Journal of Human Evolution* 39: 327–44.

Plavcan, J. M. 2001. "Sexual Dimorphism in Primate Evolution." *Yearbook of Physical Anthropology* 44: 25–53.

Plavcan, J. M. 2012. "A Re-Analysis of Sex Differences in Landscape Use in Early Hominins: A Comment on Copeland and Colleagues." *Journal of Human Evolution*, in press.

Plavcan, J. M., and Schaik, C. P. 1997. "Interpreting Hominid Behavior on the Basis of Sexual Dimorphism." *Journal of Human Evolution* 32: 345–74.

Playfair, J. 1802. *Illustrations of the Huttonian Theory of the Earth*. London: Cadell and Davies.

Potts, R. 1988. *Early Hominid Activities at Olduvai*. New York: Aldine.

Price, T. D., and Bar-Yosef, O. 2011. "The Origins of Agriculture: New Data, New Ideas." *Current Anthropology* 52 (Supplement 4): S163–74.

Pruetz, J. D., and Bertolani, P. 2007. "Savanna Chimpanzees, *Pan troglodytes verus*, Hunt with Tools." *Current Biology* 17: 1–6.

Pusey, A. E., and Packer, C. 1987. "Dispersal and Philopatry." In *Primate Societies*, edited by B. B. Smuts et al., 250–66. Chicago: University of Chicago Press.

Raichlen, D. A., Pontzer, H., and Sockol, M. D. 2008. "The Laetoli Footprints and Early Hominin Locomotor Kinematics." *Journal of Human Evolution* 54: 112–17.

Raichlen, D. A. et al. 2010. "Laetoli Footprints Preserve Earliest Direct Evidence of Human-Like Bipedal Biomechanics." *PLoS ONE* 5: e9769.

Rak, Y. 1983. *The Australopithecine Face*. New York: Academic Press.

Ramirez Rossi, F. V. et al. 2009. "Cut Marked Human Remains Bearing Neandertal Features and Modern Human Remains Associated with the Aurignacian at Le Rois." *Journal of Anthropological Science* 87: 153–85.

Read, C. 1925. *The Origin of Man*. Cambridge, UK: Cambridge University Press.

Reader, J. 1981. *Missing Links: The Hunt for Earliest Man*. London: Collins.

Reed, K. E. 1997. "Early Hominid Evolution and Ecological Change through the African Plio-Pleistocene." *Journal of Human Evolution* 32: 289–322.

Reno, P. L. et al. 2003. "Sexual Dimorphism in *Australopithecus afarensis* Was Similar to That of Modern Humans." *Proceedings of the National Academy of Science (USA)* 100: 9404–9.

Reno, P. L. et al. 2010. "An Enlarged Postcranial Sample Confirms *Australopithecus afarensis* Dimorphism Was Similar to Modern Humans." *Philosophical Transactions of the Royal Society (London) Series B* 365: 3355–63.

Rhodes, J. A., and Churchill, S. E. 2009. "Throwing in the Middle and Upper Paleolithic: Inferences from an Analysis of Humeral Retroversion." *Journal of Human Evolution* 56: 1–10.

Richerson, P. 2011. "Evolution: Not So Selfish." *Nature* 476: 29–30.

Rightmire, P., Lordkipanidze, D., and Vekua, A. A. 2006. "Anatomical Descriptions, Comparative Studies and Evolutionary Significance of the Hominin Skulls from Dmanisi, Republic of Georgia." *Journal of Human Evolution* 50: 115–41.

Ripamonti, U. et al. 1997. "Further Evidence of Periodontal Bone Pathology in a Juvenile Specimen of *Australopithecus africanus* from Sterkfontein, South Africa." *South African Journal of Science* 93: 177–78.

Robbins, M. et al. 2004. "Social Structure and Life-History Patterns in Western Gorillas (*Gorilla gorilla gorilla*)." *American Journal of Primatology* 64: 145–59.

Robinson, J. T. 1952. "Some Hominid Features of the Ape-Man Dentition." *Journal of the Dental Association of South Africa* 7: 102–13.

Robinson, J. T. 1954. "Prehominid Dentition and Hominid Evolution." *Evolution* 8: 324–34.

Robinson, J. T. 1956. *The Dentition of the Australopithecinae*. Pretoria: Transvaal Museum.

Robinson, J. T. 1961. "The Australopithecines and Their Bearing on the Origin of Man and of Stone-Tool-Making." *South African Journal of Science* 57: 3–13.

Robinson, J. T. 1962. "The Origin and Adaptive Radiation of the Australopithecines." In *Evolution und Hominisation*, edited by G. Kurth, 120–40. Stuttgart: Fischer.

Robinson, J. T. 1963. "Adaptive Radiation in the Australopithecines and the Origin of Man." In *African Ecology and Human Evolution*, edited by F. C. Howell and H. Bourliere, 385–416. Chicago: Aldine.

Robinson, J. T. 1972. *Early Hominid Posture and Locomotion*. Chicago: University of Chicago Press.

Roebroeks, W., and Villa, P. 2011. "On the Earliest Evidence for the Habitual Use of Fire in Europe." *Proceedings of the National Academy of Sciences (USA)* 108: 5209–14.

Rothschild, B. M., Herskovitz, I., and Rothschild, C. 1995. "Origins of Yaws in the Pleistocene." *Nature* 378: 343–44.

Rubidge, B. S. 2000. "Charles Kimberlin (Bob) Brain—A Tribute." *Palaeontologia Africana* 36: 1–9.

Ruff, C. 2010. "Body Size and Body Shape in Early Hominins—Implications of the Gona Pelvis." *Journal of Human Evolution* 58: 166–78.

Ruff, C. B., Trinkaus, E., and Holliday, T. W. 1997. "Body Mass and Encephalization in Pleistocene *Homo*." *Nature* 387: 173–76.

Russell, M. 2003. *Piltdown Man: The Secret Life of Charles Dawson and the World's Greatest Archaeological Hoax*. Stroud, UK: Tempus Publishing.

Ruvolo, M. 1997. "Molecular Phylogeny of the Hominoids: Inferences from Multiple Independent DNA Sequence Data Sets." *Molecular Biology and Evolution* 14: 248–65.

Ryan, A. S. 1997. "Iron-Deficiency Anemia in Infant Development: Implications for Growth, Cognitive Development, Resistance to Infection, and Iron Supplementation." *Yearbook of Physical Anthropology* 40: 25–62.

Sanders, W. J., Trapani, J., and Mitani, J. C. 2003. "Taphonomic Aspects of Crowned Hawk-Eagle Predation on Monkeys." *Journal of Human Evolution* 44: 87–105.

Sanz, C., Morgan, D., and Gulick, S. 2004. "New Insights into Chimpanzees, Tools, and Termites from the Congo Basin." *American Naturalist* 164: 567–81.

Sarmiento, E. E. 1987. "The Phylogenetic Position of *Oreopithecus* and Its Significance in the Origin of the Hominoidea." *American Museum Novitates* 2881.

Sarmiento, E. E. 1995. "Cautious Climbing and Folivory: A Model of Hominoid Differentiation." *Human Evolution* 10: 289–321.

Sarmiento, E. E. 1998. "Generalized Quadrupeds, Committed Bipeds, and the Shift to Open Habitats: An Evolutionary Model of Hominid Divergence." *American Museum Novitates* 3250.

Sarmiento, E. E. 2000. "Letter to the Editor." *Evolutionary Anthropology* 10: 15.

Sarmiento, E. E. 2010. "Comment on the Paleobiology and Classification of *Ardipithecus ramidus*." *Nature* 328: 1105b.

Savage-Rumbaugh, E. S., and Lewin, R. 1994. *Kanzi: The Ape at the Brink of the Human Mind*. New York: Wiley.

Savolainen, P. et al. 2002. "Genetic Evidence for an East Asian Origin of Domestic Dogs." *Science* 298: 1610–13.

Schaller, G. B. 1972. *The Serengeti Lion*. Chicago: University of Chicago Press.

Scheuer, L., and Black, S. 2004. *The Juvenile Skeleton*. San Diego: Academic Press.

Schick, K. D., and Toth, N. 1993. *Making Silent Stones Speak: Human Evolution and the Dawn of Technology*. New York: Simon and Schuster.

Schick, K. D. et al. 1999. "Continuing Investigations into the Stone Tool-Making and Tool-Using Capabilities of a Bonobo (*Pan paniscus*)." *Journal of Archaeological* 26: 821–32.

Schmid, P. 1983. "Rekonstruction des Skelettes bon A.L.288–1 (Hadar) und deren Konsequenzen." *Folia Primatologica* 40: 283–306.

Schmitt, D. O., Churchill, S. E., and Hylander, W. L. 2003. "Experimental Evidence Concerning Spear Use in Neandertals and Early Modern Humans." *Journal of Archaeological Science* 30: 103–14.

Schoeninger, M. J. et al. 2001. "Composition of Tubers Used by Hadza Foragers of Tanzania." *Journal of Food Composition and Analysis* 14: 15–25.

Scott, J. E., and Marean, C. W. 2009. "Paleolithic Hominin Remains from Eshkaft-e Gavi (Southern Zagros Mountains, Iran): Description, Affinities, and Evidence for Butchery." *Journal of Human Evolution* 57: 248–59.

Scott, R. S. et al. 2005. "Dental Microwear Texture Analysis Reflects Diets of Living Primates and Fossil Hominins." *Nature* 436: 693–95.

Selvaggio, M. M. 1994. "Carnivore Tooth Marks and Stone Tool Butchery Marks on Scavenged Bones: Archaeological Implications." *Journal of Human Evolution* 27: 215–28.

Selvaggio, M. M. 1998. "Evidence for a Three-Stage Sequence of Hominid and Carnivore Involvement with Long Bones at FLK *Zinjanthropus,* Olduvai Gorge, Tanzania." *Journal of Archaeological Science* 25: 191–202.

Semaw, S. et al. 1997. "2.5 Million-Year-Old Stone Tools from Gona, Ethiopia." *Nature* 385: 333–38.

Semaw, S. et al. 2003. "2.6-Million-Year-Old Stone Tools and Associated Bones from OGS-6 and OGS-7, Gona, Afar, Ethiopia." *Journal of Human Evolution* 45: 169–77.

Semaw, S. et al. 2005. "Early Pliocene Hominids from Gona, Ethiopia." *Nature* 433: 301–5.

Senut, B. et al. 2001. "First Hominid from the Miocene (Lukeino Formation, Kenya)." *Comptes Rendus de l'Académie de Sciences* 332: 137–44.

Serre, D. et al. 2006. "No Evidence of Neandertal mtDNA Contribution to Early Modern Humans." In *Early Modern Humans at the Moravian Gate,* edited by M. Teschler-Nicola, 491–504. Wien: Springer-Verlag.

Shea, J. J. 1997. "Middle Paleolithic Spear Point Technology." In *Projectile Technology,* edited by H. Knecht, 79–106. New York: Plenum Press.

Shea, J. J. 2006. "The Origins of Lithic Projectile Point Technology: Evidence from Africa, the Levant, and Europe." *Journal of Archaeological Science* 33: 823–46.

Shea, J. J., Davis, Z., and Brown, K. 2001. "Experimental Tests of Middle Paleolithic Spear Points Using a Calibrated Crossbow." *Journal of Archaeological Science* 28: 807–16.

Shermer, M. 2004. *The Science of Good and Evil: Why People Cheat, Gossip, Care, Share, and Follow the Golden Rule.* New York: Henry Holt.

Shipman, P. 1981. *Life History of a Fossil: An Introduction to Taphonomy and Paleoecology.* Cambridge, MA: Harvard University Press.

Shipman, P. 2001. *The Man Who Found the Missing Link: The Extraordinary Life of Eugène Dubois.* New York: Simon and Schuster.

Shipman, P., and Rose, J. 1983a. "Early Hominid Hunting, Butchering, and Carcass-Processing Behaviors: Approaches to the Fossil Record." *Journal of Anthropological Archaeology* 2: 57–98.

Shipman, P., and Rose, J. 1983b. "Evidence of Butchery Activities at Torralba and Ambrona: An Evaluation Using Microscopic Techniques." *Journal of Archaeological Science* 10: 465–74.

Shipman, P., and Rose, J. 1984. "Cutmark Mimics on Modern and Fossil Bovid Bones." *Current Anthropology* 25: 116–17.

Shipman, P., and Trinkaus, E. 1993. *The Neandertals: Changing the Image of Mankind.* New York: Knopf.

Shipman, P., and Walker, A. 1989. "The Costs of Becoming a Predator." *Journal of Human Evolution* 18: 373–92.

Short, L., Horne, J., and Gilbert, A. E. 2002. *Toucans, Barbets and Honeyguides.* Oxford, UK: Oxford University Press.

Shostak, M. 2000. *Nisa: The Life and Words of a !Kung Woman.* Cambridge, MA: Harvard University Press.

Shreeve, J. 1996. *The Neandertal Enigma: Solving the Mystery of Modern Human Origins.* New York: Harper Perennial.

Sikes, N. 1994. "Early Hominid Habitat Preferences in East Africa: Paleosol Carbonate Isotopic Evidence." *Journal of Human Evolution* 27: 25–45.

Simpson, G. G. 1970. "Uniformitarianism: An Inquiry into Principle, Theory, and Method in Geohistory and Biohistory." In *Essays in Evolution and Genetics in Honor of Theodosius Dobzhansky*, edited by M. K. Hecht and W. C. Steere, 43–96. New York: Appleton.

Simpson, S. W. et al. 2008. "A Female *Homo erectus* Pelvis from Gona, Ethiopia." *Science* 322: 1089–92.

Skinner, M. F. 1991. "Bee Brood Consumption: An Alternative Explanation for Hypervitaminosis A in KNM-ER 1808 (*Homo erectus*) from Koobi Fora, Kenya." *Journal of Human Evolution* 20: 493–503.

Skinner, M. F., and Goodman, A. H. 1992. "Anthropological Uses of Developmental Defects in Enamel." In *Skeletal Biology of Past Peoples: Research Methods*, edited by S. R. Saunders and M. A. Katzenberg, 153–74. New York: Wiley-Liss.

Smith, B. H. 1986. "Dental Development in *Australopithecus* and Early *Homo*." *Nature* 323: 327–30.

Smith, B. H. 1992. "Life History and the Evolution of Human Maturation." *Evolutionary Anthropology* 1: 134–42.

Smith, B. H. 1993. "The Physiological Age of KNM-WT 15000." In *The Nariokotome* Homo erectus *Skeleton*, edited by A. Walker and R. E. F. Leakey, 195–220. Cambridge, MA: Harvard University Press.

Smith, S. L. 2004. "Skeletal Age, Dental Age, and the Maturation of KNM-WT 15000." *American Journal of Physical Anthropology* 125: 105–20.

Spencer, F. 1990. *Piltdown: A Scientific Forgery*. Oxford, UK: Oxford University Press.

Spencer, F. (editor). 1997. *History of Physical Anthropology: An Encyclopedia*. New York: Garland.

Sponheimer, M., and Lee-Thorp, J. A. 1999. "Isotopic Evidence for the Diet of an Early Hominid, *Australopithecus africanus*." *Science* 283: 368–70.

Sponheimer, M. et al. 2006. "Isotopic Evidence for Dietary Flexibility in the Early Hominin *Paranthropus robustus*." *Science* 314: 980–82.

Stanford, C. B. 1998a. *Chimpanzee and Red Colobus: The Ecology of Predator and Prey*. Cambridge, MA: Harvard University Press.

Stanford, C. B. 1998b. "The Social Behavior of Chimpanzees and Bonobos—Empirical Evidence and Shifting Assumptions." *Current Anthropology* 39: 399–420.

Stanford, C. B. 1999. *The Hunting Apes: Meat Eating and the Origins of Human Behavior*. Princeton, NJ: Princeton University Press.

Stanford, C. B. et al. 1994. "Patterns of Predation by Chimpanzees on Red Colobus Monkeys in Gombe National Park, Tanzania, 1982–1991." *American Journal of Physical Anthropology* 94: 213–28.

Stanford, D. 1979. "Bison Kill by Ice Age Hunters." *National Geographic* 155: 114–19.

Stanley, S. M. 1999. *Earth System History*, 2nd ed. New York: Macmillan.

Stedman, H. H. et al. 2004. "Myosin Gene Mutation Correlates with Anatomical Changes in the Human Lineage." *Nature* 428: 415–18.

Steele, T. E., and Weaver, T. D. 2002. "The Modified Triangular Graph: A Refined Method for Comparing Mortality Profiles in Archaeological Samples." *Journal of Archaeological Science* 29: 317–22.

Steguweit, L. 1999. "Die Recken von Schöningen—400,000 Jahre Jagd mit dem Speer." *Mitteilungen der Berliner Gesellschaft für Anthropologie, Ethnologie und Urgeschichte* 8: 5–14.

Stephenson, J. 2000. *The Language of the Land: Living Among the Hazdabe in Africa*. New York: St. Martin's Press.

Stern, J. T., and Susman, R. L. 1983. "The Locomotor Anatomy of *Australopithecus afarensis*." *American Journal of Physical Anthropology* 60: 279–317.

Steudel-Numbers, K. L., and Wall-Scheffler, C. M. 2009. "Optimal Running Speed and the Evolution of Hominin Hunting Strategies." *Journal of Human Evolution* 56: 355–60.

Stiner, M. C. 1990. "The Use of Mortality Patterns in Archaeological Studies of Hominid Predatory Adaptations." *Journal of Anthropological Archaeology* 9: 305–51.

Strait, D. S., and Grine, F. E. 1999. "Cladistics and Early Hominid Phylogeny." *Science* 285: 1210.

Strait, D. S., and Grine, F. E. 2001. "The Systematics of *Australopithecus garhi*." *Ludus Vitalis* 9: 109–35.

Strait, D. S. et al. 2009. "The Feeding Biomechanics and Dietary Ecology of *Australopithecus africanus*." *Proceedings of the National Academy of Science USA* 106: 2124–29.

Stuart-Macadam, P. 1987. "Porotic Hyperostosis: New Evidence to Support the Anemia Theory." *American Journal of Physical Anthropology* 74: 521–26.

Stuart-Macadam, P. 1992. "Porotic Hyperostosis: A New Perspective." *American Journal of Physical Anthropology* 87: 39–47.

Susman, R. L. 1988. "New Postcranial Remains from Swartkrans and Their Bearing on the Functional Morphology and Behavior of *Paranthropus robustus*." In *Evolutionary History of the "Robust" Australopithecines*, edited by F. E. Grine, 149–72. New York: Aldine de Gruyter.

Susman, R. L., de Ruiter, D., and Brain, C. K. 2001. "Recently Identified Postcranial Remains of *Paranthropus* and Early *Homo* from Swartkrans Cave, South Africa." *Journal of Human Evolution* 41: 607–29.

Suwa, G. et al. 1997. "The First Skull of *Australopithecus boisei*." *Nature* 389: 489–92.

Suwa, G. et al. 2009a. "The *Ardipithecus ramidus* Skull and Its Implications for Hominid Origins." *Science* 326: 68e1–e7.

Suwa, G. et al. 2009b. "Paleobiological Implications of the *Ardipithecus ramidus* Dentition." *Science* 326: 94–99.

Symons, D. 1979. *The Evolution of Human Sexuality*. New York: Oxford University Press.

Takahata, Y., Hasegawa, T., and Nishida, T. 1984. "Chimpanzee Predation in the Mahale Moutains from August 1979 to May 1982." *International Journal of Primatology* 5: 213–23.

Tanner, N. M 1981. *On Becoming Human*. Cambridge, UK: Cambridge University Press.

Tanner, N. M., and Zilhman, A. 1976. "Women in Evolution: Innovation and Selection in Human Origins." *Signs* 1: 585–608.

Teaford, M. F., and Oven, O. J. 1989. "*In vivo* and *in vitro* Turnover in Dental Microwear." *American Journal of Physical Anthropology* 80: 447–60.

Thackeray, J. F. 1992. "On the Piltdown Joker and Accomplice: A French Connection?" *Current Anthropology* 33: 587–89.

Thackeray, J. F. 2009. "Teilhard de Chardin and 'Piltdown Man.'" *Digging Stick* 26: 17–18.

Thackeray, J. F. 2011. "On Piltdown: The Possible Roles of Teilhard de Chardin, Martin Hinton and Charles Dawson." *Transactions of the Royal Society of South Africa* 66: 9–13.

Thackeray, J. F. 2012. "Deceiver, Joker or Innocent? Teilhard de Chardin and Piltdown Man." *Antiquity* 86: 228–34.

Theunissen, B. 1989. *Eugène Dubois and the Ape-Man from Java*. Dordrecht, The Netherlands: Kluwer Academic.

Thieme, H. 1997. "Lower Paleolithic Hunting Spears from Germany." *Nature* 385: 807–10.

Tobias, P. V. 1967. *Olduvai Gorge, Vol. 2: The Cranium and Maxillary Dentition of* Australopithecus (Zinjanthropus) boisei. Cambridge, UK: Cambridge University Press.

Tobias, P. V. 1974. "Aspects of Pathology and Death among Early Hominids." *Leech* 44: 119–24.

Tobias, P. V. 1984. *Dart, Taung and the Missing Link: An Essay on the Life and Work of Emeritus Professor Raymond Dart*. Johannesburg: Wits University Press.

Tobias, P. V. 1991. *Olduvai Gorge, Vol. 4: The Skulls, Endocasts, and Teeth of Homo habilis*. Cambridge, UK: Cambridge University Press.

Tomasello, M., Call, J., and Hare, B. 2003. "Chimpanzees Understand the Psychological States of Others—The Question Is Which Ones and to What Extent." *Trends in Cognitive Sciences* 7: 153–57.

Topál, J. et al. 2009. "Differential Sensitivity to Human Communication in Dogs, Wolves, and Human Infants." *Science* 325: 1269–72.

Toth, N. 1987. "The First Technology." *Scientific American* 255: 112–21.

Toth, N. 1997. "The Artefact Assemblages in the Light of Experimental Studies." In *Koobi Fora Research Project*, Vol. 5: *Plio-Pleistocene Archaeology*, edited by G. Ll. Isaac and B. Isaac, 363–401. Oxford, UK: Oxford University Press.

Toth, N., and Schick, K. 2005. "African Origins." In *The Human Past: World Prehistory and the Development of Human Societies*, edited by C. Scarre, 46–83. London: Thames and Hudson.

Toth, N., and Schick, K. (editors). 2006. *The Oldowan: Case Studies into the Earliest Stone Age*. Bloomington, IN: Stone Age Institute Press.

Toth, N. et al. 1993. "*Pan* the Tool-Maker: Investigations into the Stone Tool-Making and Tool-Using Capabilities of a Bonobo (*Pan paniscus*)." *Journal of Archaeological Science* 20: 81–91.

Toussaint, M. et al. 2003. "The Third Partial Skeleton of a Late Pliocene Hominin (StW 431) from Sterkfontein, South Africa." *South African Journal of Science* 99: 215–23.

Trapani, J. et al. 2006. "Precision and Consistency of the Taphonomic Signature of Predation by Crowned Hawk-Eagles (*Stephanoaetus coronatus*) in Kibale National Park, Uganda." *Palaios* 21: 114–31.

Trigger, B. G. 2006. *A History of Archaeological Thought*, 2nd ed. Cambridge, UK: Cambridge University Press.

Trinkaus, E. 1983. *The Shanidar Neanderthals*. New York: Academic Press.

Trinkaus, E. 2008. "Behavioral Implications of the Muierii 1 Early Modern Human Scapula." *Annuaire Roumain d'Anthropologie* 4: 27–41.

Trinkaus, E. 2012. "Neandertals, Early Modern Humans, and Rodeo Riders." *Journal of Archaeological Science* 39: 3691–93.

Trinkaus, E., and Buzhilova, A. 2012. "The Death and Burial of Sunghir 1." *International Journal of Osteoloarchaeology*, in press.

Trinkaus, E. et al. 2006. "The Human Postcranial Remains from Mladeč." In *Early Modern Humans at the Moravian Gate: The Mladeč Caves and Their Remains*, edited by M. Teschler-Nicola, 385–445. Vienna: Springer Verlag.

Trivers, R. L. 1972. "Parental Investment and Sexual Selection." In *Sexual Selection and the Descent of Man, 1871–1971*, edited by B. G. Campbell, 136–79. Chicago: Aldine.

Trivers, R. L. 1985. *Social Evolution*. Menlo Park, CA: Benjamin/Cummings.

Trut, L. N. 1980. *The Role of Behavior in the Domestication-Associated Changes in Animals as Revealed with the Example of Silver Fox*. Ph.D. dissertation, Novosibirsk, Russia: Institute of Cytology and Genetics.

Trut, N. L. 1999. "Early Canid Domestication: The Farm Fox Experiment." *American Scientist* 87: 160–69.

Turner, A., and Antón, M. 1997. *The Big Cats and Their Fossil Relatives*. New York: Columbia University Press.

Turner, C. G. II, and Turner, J. A. 1998. *Man Corn: Cannibalism and Violence in the Prehistoric American Southwest*. Salt Lake City: University of Utah Press.

Tuttle, R. 1987. "Kinesiological Inferences and Evolutionary Implications from Laetoli Bipedal Trails G-1, G-2/3, and A." In *Laetoli: A Pliocene Site in Northern Tanzania*, edited by M. D. Leakey and J. Harris, 503–23. Oxford, UK: Clarendon Press.

Tuttle, R., Webb, D., and Tuttle, N. 1991. "Laetoli Footprint Trails and the Evolution of Hominid Bipedalism." In *Origine(s) de la bipedie chez les hominids*, edited by Y. Coppens and B. Senut, 203–18. Cahiers de Paleoanthropologie. Paris: Editions du CNRS.

Uehara, S. 1997. "Predation on Mammals by the Chimpanzee (*Pan troglodytes*)." *Primates* 38: 193–214.

Ungar, P. S. 2011. "Dental Evidence for the Diets of Plio-Pleistocene Hominins." *Yearbook of Physical Anthropology* 54: 47–62.

Ungar, P. S., Grine, F. E., and Teaford, M. F. 2006. "Diet in Early *Homo*: A Review of the Evidence and a New Model of Adaptive Versatility." *Annual Review of Anthropology* 35: 209–28.

Ungar, P. S., Grine, F. E., and Teaford, M. F. 2008. "Dental Microwear Evidence Indicates that *Paranthropus boisei* Was Not a Hard-Object Feeder." *PLoS ONE* 3, no. 4: e2044.

Ungar, P. S., and Sponheimer, M. 2011. "Early Hominin Diets." *Science* 334: 190–93.

Ungar, P. S. et al. 2012. "Dental Microwear Texture Analysis of Hominins Recovered by the Olduvai Landscape Paleoanthropology Project, 1995–2007." *Journal of Human Evolution* 63: 429–37.

van der Merwe, N. J., Masao, F. T., and Bamford, M. K. 2008. "Isotopic Evidence for Contrasting Diets of Early Hominins *Homo habilis* and *Paranthropus boisei*." *South African Journal of Science* 104: 153–55.

van der Post, L. 1980. *The Heart of the Hunter: Customs and Myths of the African Bushman*. New York: Mariner Books.

van Lawick-Goodall, J. 1968. "The Behaviour of Free-Living Chimpanzees in the Gombe Stream Reserve." *Animal Behaviour Monograph* 1: 161–311.

Vekua, A. A. et al. 2002. "A New Skull of Early *Homo* from Dmanisi, Georgia." *Science* 297: 85–89.

Vercellotti, G. et al. 2010. "Porotic Hyperostosis in a Late Upper Palaeolithic Skeleton (Villabruna 1, Italy)." *International Journal of Osteoarchaeology* 20: 358–63.

Vilà, C. et al. 1997. "Multiple and Ancient Origins of the Domestic Dog." *Science* 276: 1687–89.

Villa, P., and Soriano, S. 2010. "Hunting Weapons of Neanderthals and Early Modern Humans in South Africa: Similarities and Differences." *Journal of Anthropological Research* 66: 5–38.

Villa, P. et al. 1986. "Cannibalism in the Neolithic." *Science* 233: 431–37.

Villa, P. et al. 2009. "Stone Tools for the Hunt: Points with Impact Scars from a Middle Paleolithic Site in Southern Italy." *Journal of Archaeological Science* 36: 850–59.

Vincent, A. 1985. "Plant Foods in Savanna Environments: A Preliminary Report of Tubers Eaten by the Hadza of Northern Tanzania." *World Archaeology* 17: 131–47.

Vrba, E. S. 1981. "The Kromdraai Australopithecine Site Revisited in 1980: Recent Investigations and Results." *Annals of the Transvaal Museum* 33: 17–60.

Vrba, E. S. et al. (editors). 1995. *Paleoclimate and Evolution with Emphasis on Human Origins*. New Haven, CT: Yale University Press.

Walker, A. C. 1993. "Taphonomy." In *The Nariokotome* Homo erectus *Skeleton*, edited by A. C. Walker and R. E. F. Leakey, 40–53. Cambridge, MA: Harvard University Press.

Walker, A. C. 2009. "The Strength of Great Apes and the Speed of Humans." *Current Anthropology* 50: 229–34.

Walker, A. C., and Leakey, R. E. F. (editors). 1993. *The Nariokotome* Homo erectus *Skeleton*. Cambridge, MA: Harvard University Press.

Walker, A. C., and Shipman, P. 1996. *The Wisdom of the Bones: In Search of Human Origins*. New York: Knopf.

Walker, A. C., Zimmerman, M. R., and Leakey, R. E. F. 1982. "A Possible Case of Hypervitaminosis A in *Homo erectus*." *Nature* 296: 248–50.

Walker, A. C. et al. 1986. "2.5-Myr. *Australopithecus boisei* from West of Lake Turkana, Kenya." *Nature* 322: 517–22.

Walker, P. L. et al. 2009. "The Causes of Porotic Hyperostosis and Cribra Orbitalia: A Reappraisal of the Iron-Deficiency-Anemia Hypothesis." *American Journal of Physical Anthropology* 139: 109–25.

Wall, J. D., and Kim, S. K. 2007. "Inconsistencies in Neanderthal Genomic DNA Sequences." *PLoS Genetics* 3: 1862–66.

Wallace, J. A. 1973. "Tooth Chipping in the Australopithecines." *Nature* 244: 117–18.

Walsh, J. E. 1996. *Unraveling Piltdown: The Science Fraud of the Century and Its Solution*. New York: Random House.

Wapler, U., Crubézy, E., and Schultz, M. 2004. "Is Cribra Orbitalia Synonymous with Anemia? Analysis and Interpretation of Cranial Pathology in Sudan." *American Journal of Physical Anthropology* 123: 333–39.

Ward, C. V., Kimbel, W. H., and Johanson, D. C. 2011. "Complete Fourth Metatarsal and the Arches in the Foot of *Australopithecus afarensis*." *Science* 331: 750–53.

Ward, C. V., Leakey, M. G., and Walker, A. C. 1999. "The New Hominid Species *Australopithecus anamensis*." *Evolutionary Anthropology* 7: 197–205.

Ward, C. V. et al. 2012. "New Postcranial Fossils of *Australopithecus afarensis* from Hadar, Ethiopia (1990–2007)." *Journal of Human Evolution* 63: 1–51.

Washburn, S. L. 1957. "Australopithecines: The Hunters or the Hunted?" *American Anthropologist* 59: 612–14.

Washburn, S. L., and Avis, V. 1958. "Evolution of Human Behavior." In *Behavior and Evolution*, edited by A. Roe and G. G. Simpson, 421–36. New Haven, CT: Yale University Press.

Washburn, S. L., and Lancaster, C. K. 1968. "The Evolution of Hunting." In *Man the Hunter*, edited by R. B. Lee and I. DeVore, 293–303. Chicago: Aldine.

Watts, D. 2004. "Intracommunity Coalitionary Killing of an Adult Male Chimpanzee at Ngogo, Kibale National Park, Uganda." *International Journal of Primatology* 25: 507–21.

Watts, D., and Mitani, J. C. 2001. "Hunting and Meat Sharing by Chimpanzees at Ngogo, Kibale National Park, Uganda." In *Behavioral Diversity in Chimpanzees and Bonobos*, edited by C. Boesch, G. Hohmann, and L. Marchant, 244–55. Cambridge, UK: Cambridge University Press.

Wayne, R. K. et al. 1997. "Molecular Systematics of the Canidae." *Systematic Biology* 46: 622–53.

Welsh, F. 2000. *A History of South Africa*, revised ed. London: Harper Collins.

Wheelhouse, F. 1983. *Raymond Arthur Dart: A Pictorial Profile*. Hornsby, Australia: Transpareon.

Wheelhouse, F., and Smithford, K. S. 2001. *Dart: Scientist and Man of Gift*. Sydney: Transpareon.

White, T. D. 1978. "Early Hominid Enamel Hypoplasia." *American Journal of Physical Anthropology* 49: 79–84.

White, T. D. 1980. "Evolutionary Implications of Pliocene Hominid Footprints." *Science* 208: 175–76.

White, T. D. 1986. "Cut Marks on the Bodo Cranium: A Case of Prehistoric Defleshing." *American Journal of Physical Anthropology* 69: 503–9.

White, T. D. 1992. *Prehistoric Cannibalism at Mancos 5MTUMR-2346*. Princeton, NJ: Princeton University Press.

White, T. D. 2006. "Early Hominid Femora: The Inside Story." *Comptes Rendus Palevol* 5: 99–108.

White, T. D., Black, M. T., and Folkens, P. A. 2009. *Human Osteology*, 3rd ed. San Diego: Academic Press.

White, T. D., and Suwa, G. 1987. "Hominid Footprints at Laetoli: Facts and Interpretations." *American Journal of Physical Anthropology* 72: 485–514.

White, T. D., Suwa, G., and Asfaw, B. 1994. "*Australopithecus ramidus*: A New Species of Hominid from Aramis, Ethiopia." *Nature* 371: 306–12.

White, T. D. et al. 2006. "Asa Issie, Aramis and the Origin of *Australopithecus*." *Nature* 440: 883–89.

White, T. D. et al. 2009a. "*Ardipithecus ramidus* and the Paleobiology of Early Hominids." *Science* 326: 75–86.

White, T. D. et al. 2009b. "Macrovertebrate Paleontology and the Pliocene Habitat of *Ardipithecus ramidus*." *Science* 326: 87–93.

Whiten, A. 2005. "The Second Inheritance System of Chimpanzees and Humans." *Nature* 437: 52–55.

Whiten, A., Horner, V., and de Waal, F.B. M. 2005. "Conformity to Cultural Norms of Tool Use in Chimpanzees." *Nature* 437: 737–40.

Whiten, A., Horner, V., and Marshall-Pescinin, S. 2003. "Cultural Panthropology." *Evolutionary Anthropology* 12: 92–105.

Whiten, A., Schick, K., and Toth, N. 2009. "The Evolution and Cultural Transmission of Percussive Technology: Integrating Evidence from Palaeoanthropology and Primatology." *Journal of Human Evolution* 57: 420–37.

Whiten, A. et al. 1999. "Cultures in Chimpanzees." *Nature* 399: 682–85.

Whiten, A. et al. 2001. "Charting Cultural Variation in Chimpanzees." *Behaviour* 138: 1481–516.

Willey, G. R., and Sabloff, J. A. 1980. *A History of American Archaeology*, 2nd ed. San Francisco: W. H. Freeman.

Williams, J. et al. 2004. "Why Do Male Chimpanzees Defend a Group Range?" *Animal Behaviour* 68: 523–32.

Wilson, E. O. 1975. *Sociobiology: The New Synthesis*. Cambridge, MA: Harvard University Press.

Wilson, E. O. 1978. *On Human Nature*. Cambridge, MA: Harvard University Press.

Wilson, M., and Wrangham, R. 2003. "Intergroup Relations in Chimpanzees." *Annual Review of Anthropology* 32: 363–92.

Wobst, M. 1978. "The Archaeo-Ethology of Hunter-Gatherers or the Tyranny of the Ethnographic Record in Archaeology." *American Antiquity* 43: 303–9.

WoldeGabriel, G. et al. 1994. "Ecological and Temporal Placement of Early Pliocene Hominids at Aramis, Ethiopia." *Nature* 371: 330–33.

WoldeGabriel, G. et al. 2001. "Geology and Palaeontology of the Late Miocene Middle Awash Valley, Afar Rift, Ethiopia." *Nature* 412: 175–78.

WoldeGabriel, G. et al. 2009. "The Geological, Isotopic, Botanical, Invertebrate, and Lower Vertebrate Surroundings of *Ardipithecus ramidus*." *Science* 326: 65e1–e5.

Wolpoff, M. H. 1968. "'Telanthropus' and the Single Species Hypothesis." *American Anthropologist* 70: 477–93.

Wolpoff, M. H. 1971. "Competitive Exclusion among Lower Pleistocene Hominids: The Single Species Hypothesis." *Man* 6: 601–14.

Wolpoff, M. H. 1976. "Data and Theory in Paleoanthropological Controversies." *American Anthropologist* 78: 94–96.

Wolpoff, M. H. 1978. "Analogies and Interpretation in Paleoanthropology." In *Early Hominids of Africa*, edited by C. J. Jolly, 461–503. New York: St. Martin's Press.

Wood, B. A. 1991. *Koobi Fora Research Project, Vol. 4: Hominid Cranial Remains*. Oxford, UK: Claredon Press.

Wood, B. A. 1992. "Origin and Evolution of the Genus *Homo*." *Nature* 355: 783–90.

Wood, B. A. 1993. "Early *Homo*: How Many Species?" In *Species, Species Concepts, and Primate Evolution*, edited by W. H. Kimbel and L. B. Martin, 485–522. New York: Alan R. Liss.

Wood, B. A., and Collard, M. 1999. "The Human Genus." *Science* 284: 65–71.

Wood, B. A., and Constantino, P. 2009. "*Paranthropus boisei*: Fifty Years of Fossil Evidence and Analysis." *Yearbook of Physical Anthropology* 50: 106–32.

Wood, B. A., and Harrison, T. 2011. "The Evolutionary Context of the First Hominins." *Nature* 470: 347–52.

Wood, B. A., and Leakey, M. G. 2011. "The Omo-Turkana Basin Fossil Hominins and Their Contribution to Our Understanding of Human Evolution in Africa." *Evolutionary Anthropology* 20: 264–92.

Wood, B. A., and Lonergan, N. 2008. "The Hominin Fossil Record: Taxa, Grades and Clades." *Journal of Anatomy* 212: 354–76.

Wood, B. A., and Strait, D. S. 2004. "Patterns of Resource Use in Early *Homo* and *Paranthropus*." *Journal of Human Evolution* 46: 119–62.

Woodburn, J. 1968. "An Introduction to Hadza Ecology." In *Man the Hunter*, edited by R. B. Lee and I. DeVore, 49–55. Chicago: Aldine.

Woodburn, J. 1970. *Hunters and Gatherers: The Material Culture of the Nomadic Hadza*. London: British Museum.

Wrangham, R. W. 1986. "Ecology and Social Relationships in Two Species of Chimpanzee." In *Ecological Aspects of Social Evolution*, edited by D. L. Rubenstein and R. W. Wrangham, 352–78. Princeton, NJ: Princeton University Press.

Wrangham, R. W. 1987. "The Significance of African Apes for Reconstructing Human Social Evolution." In *The Evolution of Human Behavior: Primate Models*, edited by W. G. Kinzey, 51–71. Albany: State University of New York Press.

Wrangham, R. W. 1999. "The Evolution of Coalitionary Killing." *Yearbook of Physical Anthropology* 42: 1–30.

Wrangham, R. W., and Peterson, D. 1996. *Demonic Males: Apes and the Origins of Human Violence*. Boston: Houghton Mifflin.

Wrangham, R. W., Wilson, M., and Muller, M. 2006. "Comparative Rates of Violence in Chimpanzees and Humans." *Primates* 47: 14–26.

Yeakel, J. D. et al. 2007. "The Isotopic Ecology of African Mole Rats Informs Hypotheses on the Evolution of Human Diet." *Proceedings of the Royal Society (London) Series B* 274: 1723–30.

Zihlman, A. et al. 1978. "Pygmy Chimpanzee as a Possible Prototype for the Common Ancestor of Humans, Chimpanzees and Gorillas." *Nature* 275: 744–46.

Zipfel, B. et al. 2011. "The Foot and Ankle of *Australopithecus sediba*." *Science* 333: 1417–20.

Zollikofer, C. P. E. et al. 2002. "Evidence for Interpersonal Violence in the St. Césaire Neanderthal." *Proceedings of the National Academy of Sciences (USA)* 99: 6444–48.

Zuckerman, S. 1950. "Taxonomy and Human Evolution." *Biological Reviews* 25: 435–85.

Zuckerman, S. 1954. "Correlation of Change in the Evolution of Higher Primates." In *Evolution as a Process*, edited by J. Huxley, A. C. Hardy, and E. B. Ford, 300–352. London: Allen and Unwin.

Zuckerman, S. 1978. *From Apes to Warlords: The Autobiography (1904–1946) of Solly Zuckerman*. London: Collins.

Zuckerman, S. 1988. *Monkeys, Missiles and Men: The Autobiography (1946–1988) of Solly Zuckerman*. London: Collins.

# Index

acacia trees, 84
actualism: application of, 68–70, 83, 88–89, 100–101, 109–110; definition of, 67; "frames of plausibility" and, 85, 92, 93, 96–98, 101, 110;
Age of Reason, 126
aggression. *See* interpersonal aggression
aggressive scavenging, 74, 99, 108. *See also* hunting
Aiello, Leslie, 17
anemia, 82–83
antelope mortality profiles, 101–102
ape-man. See *Australopithecus*
"Ardi" *Ardipithecus* skeleton, 26–27. See also *Ardipithecus*
*Ardipithecus* (*kadabba* and *ramidus*): body of, 26–27, 30; brain of 10; critiques of taxonomic status of, 63; diet of, 27; discovery of, 25–26; evolutionary relationships of, 26, 28; geological age of, 10, 25–26; movement of, 26–27, 30; predation on, 120; social organization of, 28, 62–63, 113, 123; teeth of, 27–28, 61, 123
Ardrey, Robert, 45, 46, 73
arrows. *See* bows-and-arrows
atlatls, 86, 88, 91, 93
australopithecine. See *Australopithecus*
*Australopithecus aethiopicus*, 31–32
*Australopithecus afarensis*, 18, 29–30, 106–107, 109–110, 120–121

*Australopithecus africanus*, 29, 32–35, 106–107, 114, 115, 120–121; discovery of, 25, 43; "killer ape hypothesis" and, 43–45; rejection and eventual acceptance of, 39–43
*Australopithecus anamensis*, 29, 120–121
*Australopithecus:* body of, 17–18, 25, 29–30, 111–113; brain of, 10, 29; claims of butchery by, 109–110; claims of cannibalism on, 113–120; diet of, 32–39, 79, 109–110; evolutionary relationships of, 29, 31–32; geological age of, 10, 18, 25, 29, 31–32, 37, 109; movement of, 30; predation on, 68–70, 120–122; skull of, 31, 36–37; social organization of, 43–45; teeth of, 30, 33, 36–38
*Australopithecus boisei*, 37–38
*Australopithecus garhi*, 32, 35, 107–108, 109, 114–115, 123
*Australopithecus prometheus*, 30, 43–45, 114
*Australopithecus robustus*, 29–30, 32–33, 34–35, 37, 41, 43, 113, 120
autotrophy, 5

Barba, Rebeca, 82
Begun, David, 63
Berger, Lee, 121–122
Berger, Thomas, 88–89, 96
Binford, Lewis, 74–76
biochemical erosion of bone, 82